Farming Transformed in Anglo-Saxon England

Agriculture in the Long Eighth Century

Mark McKerracher

WIND*gather*
PRESS

Windgather Press is an imprint of Oxbow Books

Published in the United Kingdom in 2018 by
OXBOW BOOKS
The Old Music Hall, 106–108 Cowley Road, Oxford, OX4 1JE

and in the United States by
OXBOW BOOKS
1950 Lawrence Road, Havertown, PA 19083

Paperback Edition: ISBN 978-1-91118-831-5
Digital Edition: ISBN 978-1-91118-832-2 (epub)

A CIP record for this book is available from the British Library

Typeset in India by Lapiz Digital Services, Chennai

For a complete list of Windgather titles, please contact:

United Kingdom
OXBOW BOOKS
Telephone (01865) 241249
Fax (01865) 794449
Email: oxbow@oxbowbooks.com
www.oxbowbooks.com

United States of America
OXBOW BOOKS
Telephone (800) 791-9354
Fax (610) 853-9146
Email: queries@casemateacademic.com
www.casemateacademic.com/oxbow

Oxbow Books is part of the Casemate group

Cover images: Farmland on Magdalen Hill Down, Hampshire © Mark McKerracher

Contents

List of figures

Unless otherwise stated in caption, all maps contain OS data © Crown copyright and database right 2017

List of tables

Abbreviations

AOD   Above Ordnance Datum (measure of topographical elevation in the Appendix)

EHD   *English Historical Documents* Volume I (Whitelock 1979)

HAB   Bede, *Historia Abbatum*

HE   Bede, *Historia Ecclesiastica Gentis Anglorum*

Acknowledgements

This book has its roots in my doctoral project 'Agricultural Development in Mid Saxon England', undertaken at the University of Oxford between 2010 and 2014, and funded by the Arts and Humanities Research Council. My sincere thanks go to my supervisors, Professors Helena Hamerow and Amy Bogaard, who have given generously of their time and expertise throughout seven years of research, from germination to fruition and beyond. I would also like to thank the examiners of my thesis, Dr Mark Gardiner and Professor Mark Robinson, for their invaluable advice and critique.

It is a pleasure to thank the many kind people who have shared their expertise and data, and facilitated access to unpublished work, including (with sincere apologies for inadvertent omissions): Trevor Ashwin, Polydora Baker, Debby Banham, Angela Batt, Ian Baxter, Paul Booth, Sarah Botfield, Stuart Boulter, Esther Cameron, Gill Campbell, Jo Caruth, Brian Clarke, Pam Crabtree, Sally Croft, Anne Davis, Denise Druce, Brian Durham, Val Fryer, Sally Gale, Dave Gilbert, Jenny Glazebrook, Jessica Grimm, Julie Hamilton, Sheila Hamilton-Dyer, Sarah Howard, Anne-Marie McCann, Maureen Mellor, Mick Monk, John Moore, Jacqui Mulville, Peter Murphy, Andrew Newton, Leonora O'Brien, Nigel Page, Ruth Pelling, Colin Pendleton, Steve Preston, Sarah Pritchard, Dale Serjeantson, Kirsty Stonell Walker, Gabor Thomas, Karen Thomas, Fay Worley and Julia Wise.

The maps in this book have been produced using two free resources: the QGIS package (http://www.qgis.org, accessed April 2017) and Ordnance Survey Open Data made available under the Open Government Licence (https://www.ordnancesurvey.co.uk/business-and-government/products/opendata-products.html, accessed April 2017).

Finally, I extend my warmest thanks to my close family – to Rachel, Mum and Dad – who have been hearing about 'the book' for far too long, and who have smiled patiently when the conversation has turned to mouldboards and manure.

CHAPTER I

The lie of the land

Farming defined the Anglo-Saxon world. For the most part, its settlements were rural, its labours agricultural. Agrarian matters pervaded law-codes, riddles, miracle stories, educational texts and the reckoning of time. Land was measured less by physical extent than by agricultural capacity. Farming fed the wealth, war, craft and culture of Anglo-Saxon society; its heart beat to agrarian rhythms.

Despite being so central to the lives of Anglo-Saxon communities, agriculture has long been peripheral to Anglo-Saxon studies. A persistent dearth of evidence has rendered farming something of a poor relation, merely an assumed backdrop to greater social, political and economic themes. In particular, the scant evidence from written sources has long failed to provide any real narrative of agricultural change across the Anglo-Saxon period, between the fifth and eleventh centuries AD. It seems improbable that so long a span could have witnessed no development in farming practices but, until recently, any such processes have remained thoroughly obscure, as Hunter Blair observed:

'Wherever we look – to livestock, to cereals, to root crops, to the orchard or to the kitchen garden, it is difficult to find any evidence, at least from the earlier part of the Anglo-Saxon period, suggesting any notable innovations in comparison with the Romano-British period.' (Hunter Blair 1977, 272–273)

Over the last 40 years, however, and especially since 1990, the situation has been radically improved by an abundant harvest of new data. These decades have witnessed an extraordinary growth in Anglo-Saxon settlement archaeology, coupled with the increasingly systematic recovery and analysis of animal bones and plant remains. The study of Anglo-Saxon agriculture no longer lies outside the realm of the archaeologist, as it did for much of the twentieth century. On the contrary, a substantial, variegated dataset is now available to the agricultural archaeologist of this period. Already these developments are bearing fruit, and recent scholarship heralds an exciting new phase of research into the early medieval countryside, with landscape and settlement research by Hamerow, Rippon and Blair; seminal work on field systems by Oosthuizen, Williamson and Hall; landmark animal bone studies by Crabtree and Holmes; and the first book-length, overarching survey of the whole topic by Banham and Faith (Hamerow 2012; Oosthuizen 2013a; Williamson 2013; Hall 2014; Crabtree 2012; Holmes 2014; Banham & Faith 2014; Rippon *et al.* 2015; Blair 2013b).

Nonetheless, to date only a very few studies have closely interrogated the wide-ranging and diffuse archaeological datasets that are now available, each focusing upon a specific category of evidence such as animal bones or field systems. This book is the first systematically to draw together the evidence of pollen, sediments, charred seeds, animal bones, watermills, corn-drying ovens, granaries and stockyards on an extensive, regional scale, weaving together multiple strands of evidence in a view of agricultural development as a whole process. It utilises and integrates a diverse body of archaeological data for the first time, in order to tell a new story of farming transformed in Anglo-Saxon England.

Traditionally, Anglo-Saxon farming has been seen as the wellspring of English agriculture, setting the pattern for a thousand years to come – but it was more important than that. This book argues that the fields, ploughs, crops and livestock of Anglo-Saxon England were important not simply as the forerunners of later rural traditions, but as vital parts of the economies, cultures and societies of early medieval Britain. It focuses in particular on changes in farming practices between the seventh and ninth centuries. This period is already well known among historians and archaeologists as the time when Anglo-Saxon kingdoms and lordship became consolidated; when towns returned to the landscape for the first time since the Roman period, accompanied by an escalation in long-distance trade and craft production; and when monasteries proliferated, made wealthy by huge grants of land. This book argues that all of these momentous trends were underpinned and powered by fundamental transformations in farming. Anglo-Saxon England first came of age in its pastures and ploughland.

Previous studies have seen in this period the foundations of medieval English wool production and wheat cultivation. This book proposes a more complex picture of regional variation and specialisation. Cereal cultivation expanded massively as crop-choices were increasingly fine-tuned to local environmental conditions. New watermills, granaries and ovens were erected to cope with, and flaunt, the fat of the land. As arable farming grew at the expense of pasture, sheep and cattle came under closer management and lived longer lives, yielding more wool, dairy goods, and traction power for ploughing. These and other innovations were concentrated at royal, aristocratic and monastic centres, placing lordship at the forefront of agricultural innovation, and farming as the force behind kingdom-formation and economic resurgence in the age of Bede.

England in the 'long eighth century'

My focus upon the seventh to ninth centuries follows current trends in Anglo-Saxon scholarship, recognising this as a time of major social and economic change, in agriculture as in other spheres. Historians of early medieval Europe now write of the 'long eighth century' – encompassing the later seventh to earlier ninth centuries – as the first recognisable period when Europe emerged from its post-Roman chrysalis in a coherent new form, newly mindful of its classical heritage but following different, post-classical trajectories (Wickham 2000, ix). North-western Europe in this period was dominated – culturally, intellectually and, in some ways, politically – by Carolingian Francia which, under Charlemagne, underwent something of a classical revival in learning and law-giving. Monastic scriptoria pursued the study and copying of ancient texts, including agronomic works, while Charlemagne's legislation included what has been described as 'an explicit agrarian policy' (Butzer 1993, 558–573).

Between the seventh and ninth centuries, Anglo-Saxon England was drawn into this world: integration into European Christendom, with its rich ecclesiastical culture and fertile intellectual climate, gave rise to monastic schools of spectacular artwork and scholarship, not to mention expanding political and mercantile horizons (Webster 2012, 69–115). Later history confirms the cultural and economic impact of these centuries upon the Anglo-Saxon kingdoms. Certainly by the ninth and tenth centuries, for instance, the wealth of England – founded ultimately in agriculture – was sufficient to attract Viking raiders and then invaders, to buy peace from the same at hefty prices, and to support a network of fortified towns (burhs) to defend the imperilled realms (Higham & Ryan 2013, 232–322). When viewed from this perspective, it is difficult to imagine English agriculture *not* developing through the long eighth century.

And yet, in some ways, this is a comparatively recent view. In the later nineteenth and earlier twentieth centuries, there arose a general consensus among historians that the earlier part of the Anglo-Saxon period – the fifth and sixth centuries AD – had witnessed the most significant innovations in early medieval farming: namely, the introduction of heavy ploughing and open field systems, as part of an integrated Germanic package which ultimately foreshadowed medieval English farming (Whitelock 1952, 14; Hoskins 1955, 55). Such a view was consistent with the prevailing culture-historical paradigms of the time, in which the spread of ideas was correlated directly with the spread of peoples. In this model, the fifth and sixth centuries were seen as a period of major population flux, with Anglo-Saxon settlers widely – sometimes violently – displacing native Britons, and transforming the empty, wilding wastes of the post-Roman countryside into something more regular, more Germanised and, eventually, English. Largely circumstantial evidence, including the distribution of Old English place-names, was cited in support of this view. Compact 'Anglo-Saxon' villages and open fields were contrasted with 'Celtic' patterns of dispersed settlement and irregular fields (Green 1885, 138, 154; Maitland 1897, 15; Gray 1915, 409–418).

Such traditional narratives have long since been disputed, for several reasons. Not least among these reasons is the decoupling of migration from innovation in archaeological theory. The movement of ideas is no longer so readily ascribed to the spread of peoples. Besides this paradigm shift, various arguments have since been made against the traditional models of utter Germanisation in post-Roman Britain, whether genetic, cultural or agrarian (Hamerow 1997). Farming between the fifth and seventh centuries is now more commonly perceived as having undergone contraction and simplification, not wholesale innovation or restructuring. By contrast, the history and archaeology of the seventh to ninth centuries – spanning the long eighth century and sometimes known to archaeologists as the Mid- or Middle Saxon period – are increasingly being deemed more consistent with a model of agricultural development (Faith 2009; Rippon 2010).

It is a key premise of this book that the seventh to ninth centuries witnessed a profound change in the relationship between people and landscape in Anglo-Saxon England. This claim can be supported even before one considers specifically the evidence for farming. First, there is the political narrative, strikingly illustrated by the archaeology of the period. The later sixth and seventh centuries encompassed the crystallisation and consolidation of the kingdoms and aristocracies which came to dominate later Saxon England. New élite identities found emphatic expression in lavishly furnished burials such as those at Sutton Hoo (Suffolk), now known as 'princely graves', and settlement complexes with great halls, such as at Cowdery's Down (Hampshire). Both are vivid testimony to lordly command of wealth, labour and natural resources in this period (Wickham 2009, 157–158; Welch 2011, 269–275; Hamerow 2012, 102).

There are important implications here for agricultural history. Not only could the greater political stability thus represented have been conducive to developments in farming, but also, more specifically, these emergent élites seem to have been closely concerned and connected with agriculture. Despite its popular image as a turbulent warrior society, Anglo-Saxon England was fundamentally and necessarily an agricultural society, whose lords depended on farmers more than fighters. Kingly concern for the productive landscape was enshrined in law-codes. The late seventh-century laws of Ine of Wessex clearly demonstrate the importance of farmland in the mind-set of Anglo-Saxon royalty, dealing with issues from the fencing or hedging of shared land, to the hiring of oxen and the shearing of sheep (EHD no. 32, §§40–69). The agrarian foundations of élite wealth are even more directly attested by extant food-rents, detailing the goods to be extracted by itinerant kings and royal retinues as they traversed their subject territories. Again, Ine's Laws include a famous example (EHD no. 32, §70.1). Such concerns are also evident in charters which recorded royal land-grants from the late seventh century onwards, first for ecclesiastical beneficiaries and then, from the later eighth century, for secular landlords too. These documents typically included an appurtenance clause, summarising the product elements of their respective estates, such as 'fields, woods, meadows, pastures, fisheries, rivers, [and] springs' (EHD no. 58; Lupoi 2007, 471–473). Moreover, the fact that these land-grants were made ostensibly in perpetuity is likely to have encouraged long-term investments in their productivity (Blair 2005, 84–88; Faith 1997, 30; Wormald 1984, 19–23).

Such material concerns were not at all chastened by the Christian conversion of the Anglo-Saxon kingdoms, nominally complete by the 680s (Blair 2005, 9–49). Land charters were originally used to endow early monastic foundations with vast productive estates. More generally, as Anglo-Saxon kings and kingdoms were drawn within the ambit of western Christendom, opportunities for cultural, intellectual and material exchanges intensified: fertile ground for agricultural innovation at élite levels.

The humbler landscapes of peasant farming, under new governance and sometimes changing hands by royal writ, were also transformed in this period, as reflected in the archaeological remains of rural settlements. There are several important differences between rural settlements of the seventh to ninth centuries and their fifth- to sixth-century precursors, which together suggest the growing importance of proprietary land-rights over time. Essentially, rural communities were becoming more closely and formally attached to their farmlands. Whereas fifth- to sixth-century settlements displayed little if any formal planning and could change location over their lifetimes, their seventh-century and later successors tended to show both greater formality and greater stability (Hamerow 2012, 22–24, 67–119; Ulmschneider 2011, 160–165). At the same time, a growing tendency for graves to be situated within or near settlements could, as Hamerow has argued, have served to legitimise ancestral claims to local land (Hamerow 2012, 129).

The wider economic picture of England in the long eighth century, meanwhile, tells of growing organisation, specialisation and intensification: a step-change in economic activity which, as I will argue, was ultimately powered by agriculture. Inland markets and coastal emporia, both entirely new kinds of settlement in this period, served as nodes in a spreading network of specialist craft and trade involving some degree of monetisation (Pestell 2011). Following a post-Roman hiatus in the fifth and sixth centuries, coinage returned to England first as gold tremisses (*c.* AD 580–675) and later, much more abundantly, as silver sceattas (*c.* AD 680–750). Another remarkable innovation of the seventh- to ninth-century economy is the mass production and wide distribution of Ipswich Ware (*c.* AD 720–850), the first wheel-turned, kiln-fired pottery made in Britain since the Roman period (Blinkhorn 2012). Other contemporary innovations in craft production include specialist iron-smelting at sites such as Ramsbury (Wiltshire), and the increased production of standardised metal dress accessories in the eighth and ninth centuries (Naylor 2012, 248–252; Thomas 2011, 412–414).

Finally, to set alongside the social, political and economic accounts, there is the evidence from archaeology's sister-subject, palaeoclimatology. Changes in the climate would no doubt have affected the parameters of agricultural production in the past, but this area can prove contentious. The chief difficulties for the archaeologist lie, first, in unravelling how climates were changing at any given point in time and, second, in interpreting the relationship between such global climatic trends and farming practices at a given locale. With regard to the first point, a wide range of evidence has been marshalled in studies of climate change across the first millennium AD, including documentary sources, pollen, ice cores and tree rings. Following the comparatively warm centuries of Roman rule in Western Europe, cooler and wetter conditions appear to have prevailed from the fifth century onwards. A recent study has also argued that the climatic instability of the fifth and sixth centuries, the disruptive rapidity of change, could itself have had adverse effects on farming (Büntgen *et al.* 2011, 578–582). Warmer and drier conditions returned in the last quarter of the millennium, continuing, albeit not uniformly, into the so-called 'Medieval Warm Period' (Hughes & Diaz 1994; Dark 2000, 19–28).

However, these are broad and debateable trends. A more detailed consensus on early medieval climate change – and its likely impact on human activity – is not forthcoming for England in the long eighth century, let alone for the smaller regions studied in this book. Such detail is perhaps beyond the reach of palaeoclimatology's global and long-term data models. Climatological proxy data cannot directly support this book's contention that English farming became markedly more productive between the seventh and ninth centuries AD. On the other hand, they do not contradict it. One can at least say that a narrative of agricultural development in seventh- to ninth-century England would be entirely compatible with current models of global climate change for the first millennium AD.

Rationale and scope of this study

The massive growth in development-led excavation in British archaeology since 1990 has brought to light an unprecedented number of Anglo-Saxon settlement sites, too many for the individual scholar adequately to track. This is the result of changes to planning policy – *PPG* 16 and its successors – making developers responsible for undertaking archaeological work on construction sites (Bradley 2006). In addition, the recovery and analysis of bioarchaeological remains from excavated sites, especially animal bones and macroscopic plant remains, has become more routine and systematic over the past 30 years. The size of

FIGURE 1. Location of the case study regions within Britain

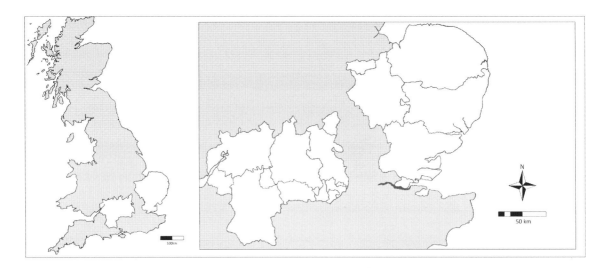

the resultant dataset is so great that it has not been practical, in this book, to attempt a comprehensive study of agricultural development across the entirety of Anglo-Saxon England. This work therefore focuses upon evidence from two extensive case study regions, sufficiently rich in data to yield meaningful results (Figure 1). To the west is a region centred on the Upper and Middle Thames valley, incorporating Gloucestershire, Wiltshire, Oxfordshire, Buckinghamshire, Berkshire, and associated unitary authorities. To the east is an East Anglian region incorporating Cambridgeshire, Norfolk, Suffolk, Essex, and associated unitary authorities (Figure 2). Excavations in these regions have discovered relatively large numbers of Anglo-Saxon settlements, in comparison with others parts of Britain. Indeed, the archaeological authenticity of a core Anglo-Saxon culture zone based around the Great Ouse and Upper Thames valleys has become increasingly convincing in recent scholarship (Hamerow 2012, 4, fig. 1.1; Blair 2013a, 5–8).

The two regions are defined by modern administrative boundaries in order to facilitate the collection of archaeological data; they are not presupposed to have any historical or environmental coherence. On the contrary, it is intended that they should embrace a variety of landscapes – such as downland, fenland, valleys and heaths – so that environmental influences upon farming practices may be explored. In addition, although the selected regions proved to be archaeologically fruitful, it was inevitable that many relevant sites would often fall outside their limits (such as Flixborough, Lincolnshire, with its large animal bone assemblage; or the diffuse corpus of Saxon watermills). Sites from elsewhere in Anglo-Saxon England have therefore been discussed where relevant,

FIGURE 2. Modern administrative composition of the case study regions

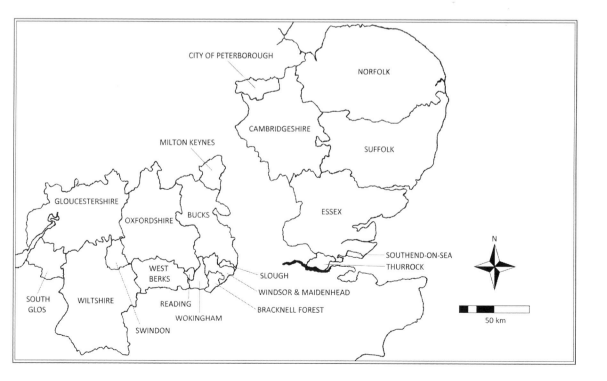

but only those from within the study regions are considered to form part of the core dataset. Hence, references in the text to the 'project dataset' exclusively denote evidence from the two study regions, specifically from the sites detailed in the Appendix.

Beating the bounds: natural environments in the study regions

It is commonplace for excavation reports to include a preliminary section describing the natural environment of a site, including notes on geology, topography, land-use and watercourses. Such accounts are easily overlooked and not always integrated with archaeological discussions of the site in question. These natural details are, however, of obvious relevance to this study. Environmental factors will have determined not only the preservation of archaeological material but also the availability of natural resources, and will thus have constrained agricultural strategies in antiquity. My use of the word 'constrained' may jar, if it is taken to mean a denial of decisive human agency. Archaeologists and historians are sometimes at pains to forestall accusations of environmental or geographic 'determinism'. But there is surely a converse risk of overlooking an important truth here, a truth recognised, perhaps, in the lengthy, detailed encomium on the natural bounty of Britain and Ireland with which Bede opens his *Historia Ecclesiastica*: human production is rooted firmly in natural ground (Bede, *H.E.* I.1). Rural communities must work with whatever nature has provided, and, if nature's provision is inadequate for their purposes, then they must work all the harder to make good these shortcomings. Either way, their activities and choices are constrained. It is only to be expected, therefore, that agricultural patterns will vary geographically as well as chronologically upon the variegated natural canvas of the study regions. I have here found inspiration in the self-proclaimed 'cow and plough' approach of Tom Williamson, to whose work much of this chapter is indebted (Williamson 2003; 2013).

So the story starts with the landscape. Ultimately, this means considering its physical structure – geology, topography, watercourses – but climate is also an important factor. While one cannot necessarily expect current climates exactly to resemble those of Anglo-Saxon England, there are broad climatic trends across the study regions which, being due to location and landform, should be just as relevant for the early medieval period as for today. Thus, a more 'Atlantic' climate is experienced in the west of the study regions, with average temperatures and rainfall both relatively high. By contrast, the more 'Continental' climate of East Anglia, exposed to the North Sea, brings lower average rainfall but greater extremes of temperature, especially on the Breckland plateau (Pryor 2010, 22; Williamson 2013, 42–43). As Williamson has argued, citing modern agronomical models, this climatic gradient is likely to have exerted a crucial influence on cereal farming, since the size and reliability of cereal yields is intimately tied to rainfall patterns. The drier eastern reaches of the study regions could therefore be expected to have enjoyed larger and more reliable harvests than the lands further west. It is chiefly to this factor that Williamson attributes the high population densities of East Anglia recorded in the Domesday Survey of 1086 (Williamson 2003, 35).

Turning now to the physical composition of the landscape, to the rocks and soils which have constituted farmland ancient and modern, it is beyond the scope of this book to attempt a detailed geological and edaphic account of the study regions. What follows is necessarily simplified but outlines the key physical characteristics, detailing some of the natural bounds within which the Anglo-Saxon famers had to forge their living.

Both study regions are situated well within Britain's Lowland zone. Their bedrock (or solid geology) tends to be soft and readily eroded – chalk, limestone, clay – so that even the uplands seldom rise much more than 200 m above sea level (Williamson 2013, 36–40). Across much of this area, the bedrock is overlain by more recent drift (or superficial) geology, largely glacial deposits of sand, gravel and clay which now characterise many river valleys in the study regions. The riverine landscapes in the clay vales of the Upper and Middle Thames tend to lie between 50 and 100 m above sea level, while glacial scouring around the Fens and Lower Thames has left much of this land even lower-lying, mostly fewer than 50 m (in the fenland, barely 5 ms) above sea level (Figure 3).

The general shape of the surface landforms is determined by bands of chalk and limestone, which effectively divide the study regions into latitudinal strips (Figure 4). A band of chalk runs south-west to north-east from Salisbury Plain to northern Norfolk, roughly dissecting the two regions. It accounts for the uplands and plateaux of Salisbury Plain, the Wessex Downs, Chilterns, East Anglian Heights and Breckland. Another, connected band of chalk running west to east, from Salisbury Plain to northern Kent, falls mainly outside the study regions and roughly demarcates their southern boundary. The northern boundary of the study regions is similarly marked by the raised limestone band of the Cotswolds, Northamptonshire uplands, and Rockingham Forest. Between these raised lands of limestone and chalk are the lower-lying clayey plains, river valleys and wetlands which define the basic hydrological patterns in this stretch of England. The Thames and its many tributaries drain most of the western study region roughly south-eastwards; likewise much of Essex draining into the Thames and Blackwater estuaries. Much of East Anglia, meanwhile, plus the area around Milton Keynes, is drained north-eastwards into the Wash by the Nene and Great Ouse and their tributaries (Williamson 2003, 56–59).

This cursory description of the regions' geology and topography cannot do justice to their great variety of landscapes, and offers no systematic means of relating sites to their natural settings. I have therefore divided the area into several sub-regions, a simplistic but useful means of classifying archaeological sites in terms of geography (Figure 5). These sub-regions are based upon the 'Natural Areas' and 'National Character Areas' devised by Natural England, which I have slightly adapted in response to the actual distribution of sites

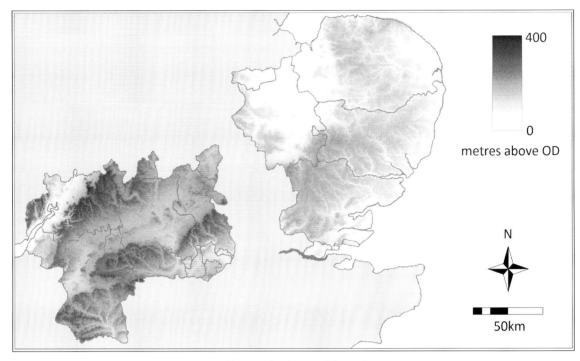

FIGURE 3. Topographical map of the study regions

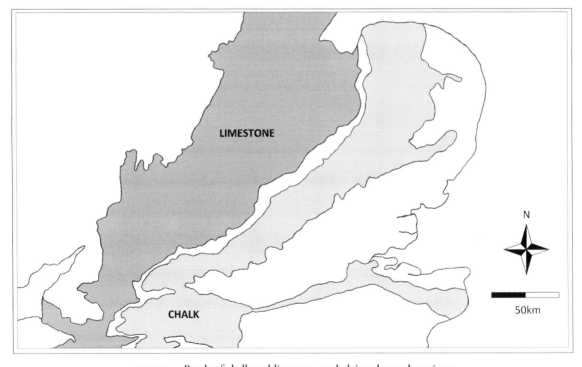

FIGURE 4. Bands of chalk and limestone underlying the study regions

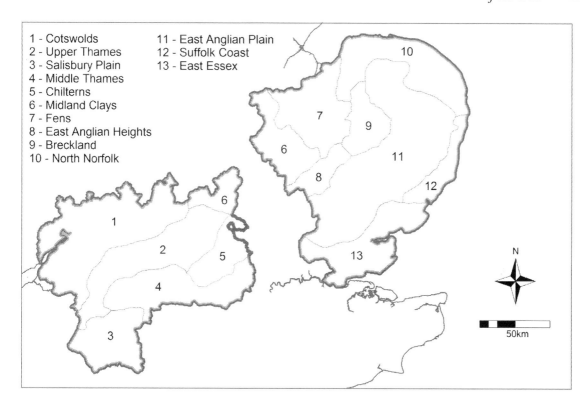

1 - Cotswolds
2 - Upper Thames
3 - Salisbury Plain
4 - Middle Thames
5 - Chilterns
6 - Midland Clays
7 - Fens
8 - East Anglian Heights
9 - Breckland
10 - North Norfolk

11 - East Anglian Plain
12 - Suffolk Coast
13 - East Essex

FIGURE 5. Geographical sub-regions used in this study

within my dataset. The sub-regions are named after their respective dominant natural features, which are not always the richest in data: most sites recorded in the Chilterns sub-region, for instance, in fact lie below the scarp, overlooking the Vale of Aylesbury (see below). And as a general observation, it should be noted that most sites lie at, or close to, the junction of two or more types of terrain; few are surrounded by a uniform landscape. The Appendix may be consulted for a site-by-site review.

Cotswolds

The highest and westernmost of the sub-regions is centred on the western reaches of the Cotswolds (Figure 6). The sheer scarp face of this limestone belt overlooks the Severn and Avon vales to the north-west, while the dipslope rolls gently south-eastwards, it many rivers feeding ultimately into the Upper Thames. Data relevant to this project proved sparse in this zone, with few sites recorded on the higher limestone uplands with their thin, brashy soils. At the escarpment, Uley alone takes the high ground of an outlying spur of Cotswold limestone. Below the scarp, overlooking the Severn and Avon vales, Bishop's Cleeve, Frocester Court and Dumbleton lie upon sands and gravels. On the Cotswold dipslope, while Barnsley Park lies upon the prevailing substrate of limestone, overlain in places with boulder clay, Lower Slaughter occupies a

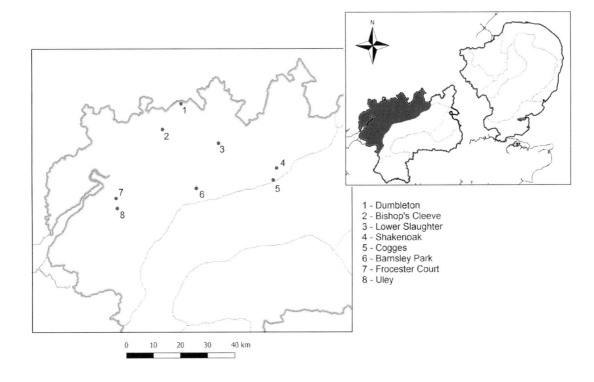

1 - Dumbleton
2 - Bishop's Cleeve
3 - Lower Slaughter
4 - Shakenoak
5 - Cogges
6 - Barnsley Park
7 - Frocester Court
8 - Uley

gravel terrace near the Slaughter Brook, one of the many streams and rivers that dissect the hills. To the south-east, where the dipslope meets the Upper Thames, limestone outcrops occur amongst the gravel terraces and clay vales: the sites at Cogges and Shakenoak are located in such situations, with access to heavy clays and well-drained gravels as well as the brashy upland soils of the Cotswold hills.

FIGURE 6. Location of sites in the Cotswolds sub-region

Upper Thames

The clay vales of the Upper Thames and its tributaries host far higher concentrations of sites in the project dataset (Figure 7). Here are combined the heavy, relatively intractable soils of the Jurassic clays, the lighter, free-draining soils of the gravel terraces, and the alluvial clays of the floodplains. The modern riverine landscape presents a much simplified picture of what, in the Anglo-Saxon period, would have been a more complex, braided network of channels. The complexity has since been diminished through canalisation and a long process of silting which resumed – after a post-Roman hiatus – in the later Anglo-Saxon period (Booth *et al.* 2007, 6, 18–20). Most of the sites recorded here are found on the gravel terraces, a trend due in part to the enhanced discovery of settlements through modern gravel quarrying, but also attributable to the relatively easily tilled soils and ready access to water in these locations. The great majority of sites studied here are situated upon either the First (Floodplain) or Second (Summertown-Radley) gravel terraces; it is simpler

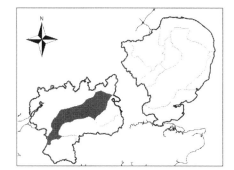

1 - Latton Quarry
2 - Lechlade
3 - New Wintles Farm
4 - Eynsham
5 - Worton
6 - Yarnton
7 - Cresswell Field
8 - Alchester
9 - St Aldate's, Oxford
10 - Rycote
11 - Littlemore
12 - Barrow Hills
13 - Barton Court Farm
14 - Spring Road, Abingdon
15 - Berinsfield
16 - Dorchester-on-Thames
17 - Benson
18 - Neptune Wood
19 - Sutton Courtenay
20 - Didcot
21 - Wantage

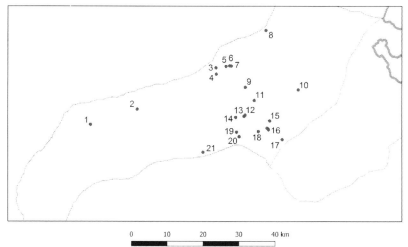

FIGURE 7. Location of sites in the Upper Thames sub-region

to name those few sites in the Upper Thames environs which do *not* occupy gravel terrace locations. There are three sites in what is now the St Aldate's area of Oxford, which in the early medieval period would have been a floodplain with seasonally flooded islands of alluvial clay (Dodd 2003, 12–16). A little to the south are the low hills of the Midvale Ridge, which rises amid the clay vales. Formed on limestone, some of its sandy soils are similar to those of the East Anglian Breckland, dry and acidic (discussed below). Among these hills are the sites at Rycote and Littlemore. Finally, away from the Thames to the south-west, lies Wantage at the interface between the clay Vale of the White Horse and the chalky head deposits at the north-facing scarp of the Lambourn Downs.

Salisbury Plain

The chalk uplands of the Lambourn Downs, along with their extensions east and west, mark the southern boundary of the Upper Thames sub-region. To the south-west, this chalk belt forms the gently rolling uplands of Salisbury Plain, noted today for its extensive calcareous grassland (Figure 8). The sites in the project dataset are located rather to the margins of the plain, near the intersecting gravelly river valleys of the Bourne, Avon, Wylye and Nadder.

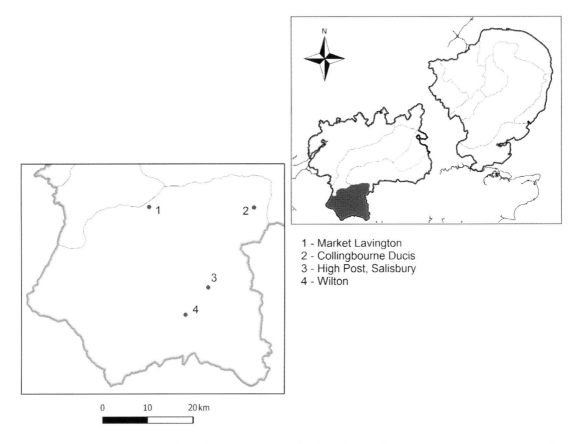

1 - Market Lavington
2 - Collingbourne Ducis
3 - High Post, Salisbury
4 - Wilton

Market Lavington is on a ridge of greensand, with the chalk plateau to the south and clays to the north; in the valley immediately below the site, around the Easterton Brook palaeochannel, are peat and alluvium. Sites at Collingbourne Ducis and High Post (Salisbury) lie upon chalk, but with river gravels nearby, while Wilton is situated on a gravel spur between the rivers Wylye and Nadder.

FIGURE 8. Location of sites in the Salisbury Plain sub-region

Middle Thames

Continuing eastwards, south of the chalk belt, we reach the Middle Thames and its wider catchment, including major tributaries such as the Kennet (Figure 9; Booth *et al.* 2007, 7). The Middle, as distinct from the Upper, Thames begins at the Goring Gap, a riverine passage that cuts through the chalk belt. Goring itself occupies a gravel terrace, as do Taplow Court, Old Windsor, Burghfield, the Dorney sites (Lot's Hole and Lake End Road), Forbury House (Reading), Ramsbury and Hungerford. It should be noted that the flinty gravel terraces of the Middle Thames valley are often more acid and less fertile than their Upper Thames counterparts. Similarly, the sites at Chieveley and Wickhams Field lie on the mottled clays and sands of the Reading Beds, which often give rise to sandy, acidic soils.

1 - Ramsbury
2 - Hungerford
3 - Chieveley
4 - Goring
5 - Wickhams Field
6 - Forbury House, Reading
7 - Taplow Court
8 - Dorney
9 - Old Windsor

0   10   20   30   40 km

FIGURE 9. Location of sites in the Middle Thames sub-region

Chilterns

Rising to the north over the Middle Thames valley is the dipslope of the Chilterns (Figure 10). The Chiltern Hills themselves, a continuation of the long chalk belt that runs through the study regions, are characterised by light, calcareous soils, and are barren of data in this project's dataset (Williamson 2003, 26–34). On the dipslope too, an area of acidic soils, pertinent data are sparse: only Latimer, to the south, occupies the chalky slopes of the Chess valley, with its gravel and alluvium deposits. More sites are found to the north, at the foot of the Chilterns escarpment where it meets the clays of the Vale of Aylesbury. Here, the Lower Icknield Way site occupies clayey ground, while Pitstone lies on chalky head deposits. West of here, a cluster of five sites – perhaps representing one extended settlement sequence – is located in the Walton area of Aylesbury, on a soft outcrop of clayey, sandy Portland limestone.

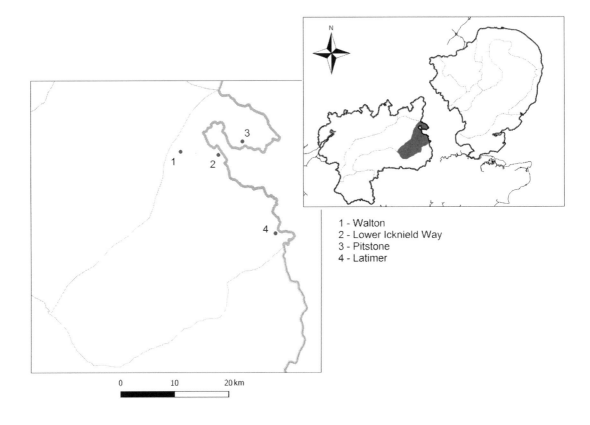

1 - Walton
2 - Lower Icknield Way
3 - Pitstone
4 - Latimer

Midland Clays

Following to the north-east, we enter the rolling landscape of the Midland Clays (Figure 11). These claylands gradually descend from south-west to north-east – spanning both study regions – to drain ultimately into the Wash via the Great Ouse, Nene, and associated river systems. This often fertile zone embraces a complex series of sands, gravels and clays, often giving rise to damp, heavy, intractable terrain but also, in the river valleys, some better-drained soils (Williamson 2003, 25–26, 63–65). The project dataset is very well represented here, and can be divided roughly between three groups of sites. First, to the north of the zone, the sites at Orton Waterville and Orton Hall Farm occupy gravel terraces in close juxtaposition with clays, while Wittering – a site with evidence of metalworking – is situated upon an outcrop of ironstone in the Rockingham Forest area. Second, to the south-west, there is a concentration of sites associated with Milton Keynes, resulting from the prolific excavations that attended the construction of the 'new town' there in the later twentieth century, as well as subsequent urban development (Zeepvat 1993). These sites are spread mostly across gravel deposits, amid the clay ridges and outcropping limestone in the area. Third, there is a central group of sites in a reasonably

FIGURE 10. Location of sites in the Chilterns sub-region

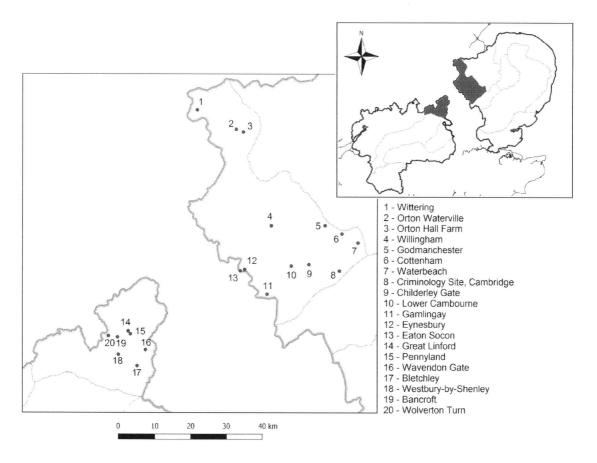

1 - Wittering
2 - Orton Waterville
3 - Orton Hall Farm
4 - Willingham
5 - Godmanchester
6 - Cottenham
7 - Waterbeach
8 - Criminology Site, Cambridge
9 - Childerley Gate
10 - Lower Cambourne
11 - Gamlingay
12 - Eynesbury
13 - Eaton Socon
14 - Great Linford
15 - Pennyland
16 - Wavendon Gate
17 - Bletchley
18 - Westbury-by-Shenley
19 - Bancroft
20 - Wolverton Turn

FIGURE 11. Location of sites in the Midland Clays sub-region

fertile zone of boulder clay plateaux (Williamson 2003, 72). Childerley Gate and Lower Cambourne occupy the boulder clay, but most of the other sites here lie upon sands and gravels in the valleys of the Cam, Great Ouse, and associated river systems. To the south of these sites, however, Gamlingay sits upon the Bedfordshire Greensand Ridge, characterised by infertile soils; and Cottenham, between Willingham and Waterbeach, similarly occupies a greensand ridge location.

Fens

Abutting the claylands to the north-east is one of the most distinctive zones within the study regions, the Fens of East Anglia (Figure 12). These vast, open wetlands have been subject to extensive drainage and reclamation in modern times. Indeed, defensive coastal banks began to be erected from the ninth or tenth century onwards, but the fifth- to eighth-century littoral would have been more mutable, and far inland of the modern coastline (Figure 13; Crowson *et al.* 2005, 4–11). The marshy, estuarine silt fens would therefore

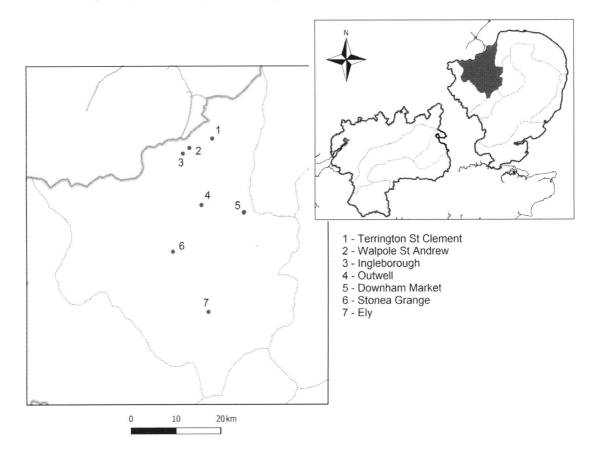

1 - Terrington St Clement
2 - Walpole St Andrew
3 - Ingleborough
4 - Outwell
5 - Downham Market
6 - Stonea Grange
7 - Ely

have been subject to repeated (if short-lived) marine transgressions from the Wash, with correspondingly saline influences on their habitats. In the project dataset, silt fen sites are recorded at Terrington, Walpole and Ingleborough; they lie upon raised beds of silt known as roddons, just three metres above sea level. Further inland are the peats of the so-called Black Fens, where again settlement was restricted to raised islands amid the wetlands: at Stonea Grange, upon sand and gravel; and at three sites in Ely, upon an outcrop of sandstone, sands and clays. Around the margin between the silt and peat fens, sites at Outwell and Downham Market occupy sands and clays.

FIGURE 12. Location of sites in the Fens sub-region

East Anglian Heights

To the south, the chalk belt continues to run west to east, through the East Anglian Heights and Breckland. First, the East Anglian Heights are relatively high ground bearing mostly thin, poor, calcareous soils (Figure 14). The higher, chalky ground is lacking in relevant data. Cherry Hinton occupies a lower-lying area of chalk, but otherwise most sites are situated upon – or close to – gravel terraces, where rivers dissect the chalk.

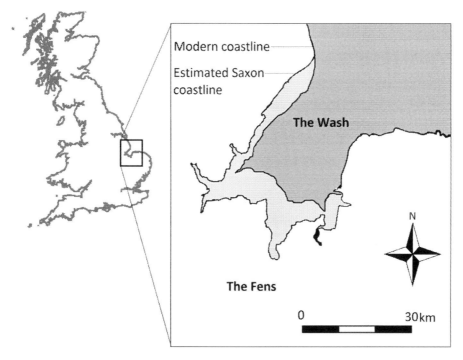

FIGURE 13. Changing coastline around the Wash and Fenland (after Crowson *et al.* 2005, 5, fig. 1)

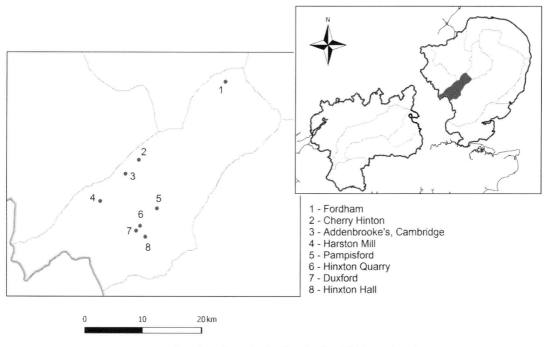

1 - Fordham
2 - Cherry Hinton
3 - Addenbrooke's, Cambridge
4 - Harston Mill
5 - Pampisford
6 - Hinxton Quarry
7 - Duxford
8 - Hinxton Hall

FIGURE 14. Location of sites in the East Anglian Heights sub-region

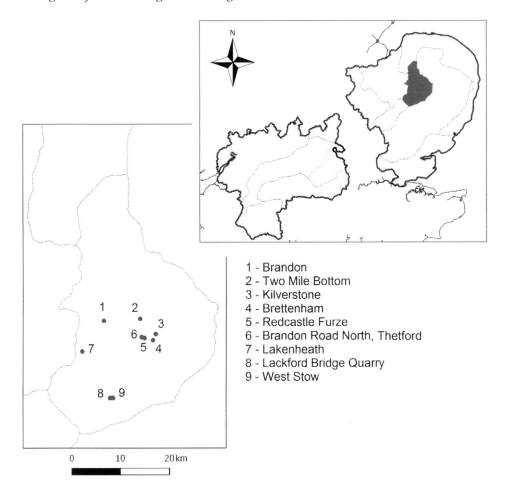

1 - Brandon
2 - Two Mile Bottom
3 - Kilverstone
4 - Brettenham
5 - Redcastle Furze
6 - Brandon Road North, Thetford
7 - Lakenheath
8 - Lackford Bridge Quarry
9 - West Stow

FIGURE 15. Location of sites in the Breckland sub-region

Breckland

The chalk upland extends further north as the low plateau of the Breckland (Figure 15), where it is largely covered in poor, dry, sandy, acid soils (Williamson 2003, 270). This warm, dry plateau today supports much heathland, and is dissected by a number of rivers. It is on the gravels and sands of these river valleys that most relevant Anglo-Saxon sites are situated: Two Mile Bottom beside the Little Ouse; West Stow and Lackford Bridge Quarry in the Lark valley; and a cluster of sites in the vicinity of Thetford, beside the Thet and Little Ouse rivers: Kilverstone, Redcastle Furze, Brandon Road North, and Brettenham. A tight cluster of sites at Lakenheath and a high-status site at Brandon both lie upon sands, but are in close proximity to the richer, peaty soils where the Breckland meets the edge of the fens.

North Norfolk

Directly to the north is a heterogeneous zone spanning North Norfolk (Figure 16). Just beyond the Breckland, Marham sits on the springline between

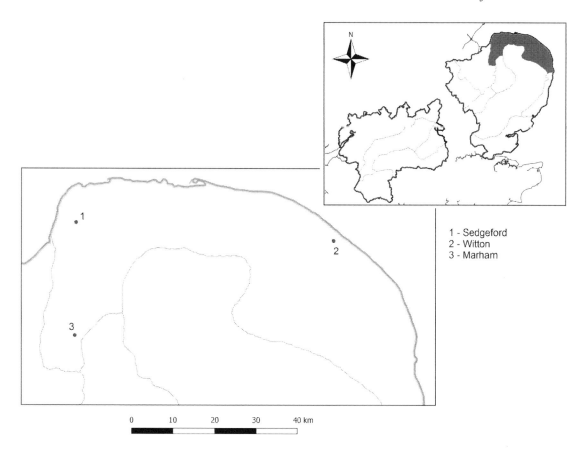

1 - Sedgeford
2 - Witton
3 - Marham

FIGURE 16. Location of sites in the North Norfolk sub-region

lower and middle chalk. To the north-west of Norfolk, calcareous lowlands lie beside more acidic sandy soils on the modest uplands. Here, Sedgeford occupies sand and gravel beside the marshy floodplain of the river Heacham. The north-east is also a sandy zone, but notably fertile in places. Here, again, Witton occupies sand and gravels (Williamson 2003, 27, 86–87).

East Anglian Plain

Moving south into the heart of East Anglia, we find the largest sub-region in this study: the East Anglian Plain, a great band of chalky boulder clay giving rise to slow-draining but often fertile clayey and loamy soils, with sands, gravels and alluvium in the river valleys (Figure 17). Such is the general character of this district but, as Williamson describes, there are important distinctions between its northern and southern portions, separated by the Gipping valley (Williamson 2003, 26–27, 94–101). In the comparatively poorly-drained northern portion of these claylands, recorded sites are largely concentrated upon (or close to) better-drained areas of sands and gravels amid the clays: thus North Elmham Park, Spong Hill, Wymondham, Flixton Park Quarry, and Eye. The southern East Anglian Plain is a more undulating landscape than that to the north, more frequently dissected by rivers with alluvial floodplains, and characterised on the whole by better-drained, calcareous soils (Williamson 2003, 101). In this zone, Wicken Bonhunt is located on gravel and sandy clay, and two sites at Stansted are situated upon the boulder clay plateau itself.

Located between these northern and southern zones, there is a tight cluster of sites in the vicinity of modern Ipswich. This is an interesting location, for it lies not only at the junction of the northern and southern portions of the East Anglian Plain, but also in relatively close proximity to the sandy coast of Suffolk (discussed below). Four sites in the project dataset are located on the clays, sands and gravels that converge in this riverine location, around the modern city: Whitehouse Road, Handford Road, Hintlesham, and central Ipswich. The latter 'Ipswich' is here used as a convenient shorthand for what is in fact a composite site, comprising several different rescue excavations undertaken in the urban core of modern Ipswich since 1974, representing the emporium of Gipeswic which flourished in the long eighth century. These excavations have not been fully published, individually or collectively, but available reports on the animal and plant remains treat central Ipswich as a whole in this composite fashion, and accordingly I have adopted the same approach (Crabtree 2012, 5).

FIGURE 17. Location of sites in the East Anglian Plain sub-region

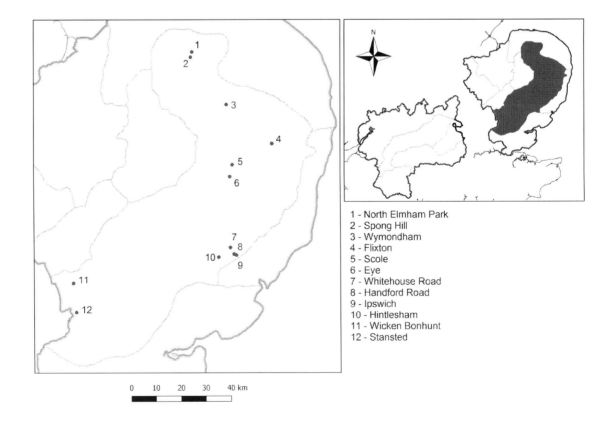

1 - North Elmham Park
2 - Spong Hill
3 - Wymondham
4 - Flixton
5 - Scole
6 - Eye
7 - Whitehouse Road
8 - Handford Road
9 - Ipswich
10 - Hintlesham
11 - Wicken Bonhunt
12 - Stansted

0   10   20   30   40 km

Suffolk Coast

To the east of the East Anglian Plain, the project dataset includes only a single site along the Suffolk Coast (Figure 18), a low-lying zone known locally as the Sandlings, chiefly characterized by sandy heathland and low average rainfall. The soils here are light and easily tilled, but free-draining, acidic and infertile. The site at Bloodmoor Hill, however, is situated close to a clay ridge extending from the East Anglian clays to the west, so its agricultural activities need not have been restricted to the sands (Williamson 2003, 95–96; Lucy *et al.* 2009, 1–4).

East Essex

FIGURE 18. Location of sites in the Suffolk Coastal Strip sub-region

Finally, in the south-eastern corner of the study regions, the relatively fertile claylands of the East Anglian Plain give way to the London Clay soils of southern Essex: infertile, acidic, intractable and prone to waterlogging. As Williamson observes, this naturally 'inhospitable' and 'uninviting' terrain remained relatively

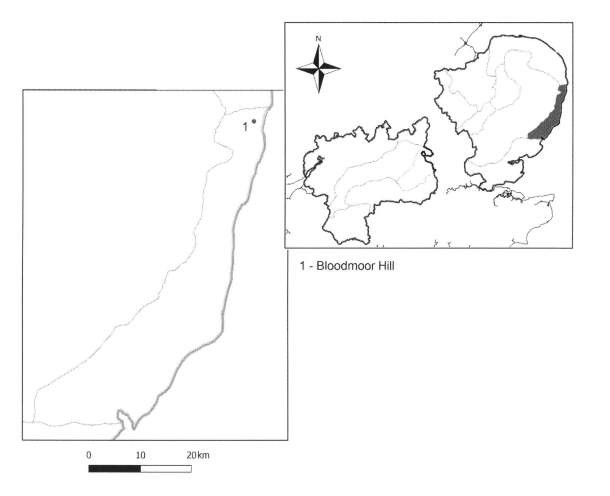

1 - Bloodmoor Hill

0   10   20km

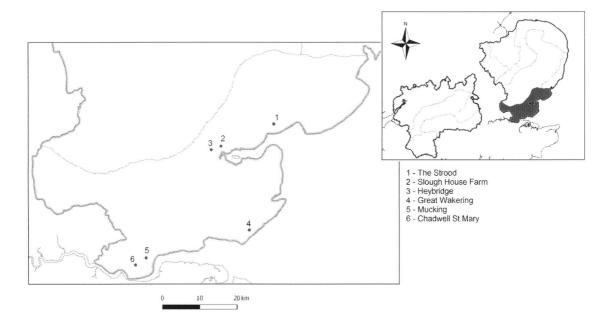

1 - The Strood
2 - Slough House Farm
3 - Heybridge
4 - Great Wakering
5 - Mucking
6 - Chadwell St Mary

heavily wooded into the later Anglo-Saxon period, according to the testimony of the Domesday Book, Old English place-names and charters (Rackham 1986, 75–85; Williamson 2003, 51–60; Roberts & Wrathmell 2000, 31, fig. 24). So it is perhaps unsurprising that the sites in the project dataset are largely peripheral to such poor and wooded terrain, instead occupying riverine, estuarine and coastal locations in an East Essex zone (Figure 19). Mucking, Chadwell St Mary, Slough House Farm and Heybridge are all situated on river terrace gravels, while Great Wakering occupies brickearth overlying a gravel terrace.

And so the bounds of this study have been beaten. This chapter has outlined the historical, archaeological and geographical contexts of the project, providing an essential framework for the analyses and discussions which follow.

FIGURE 19. Location of sites in the East Essex sub-region

CHAPTER 2

Farm and field

Laxton is an anomaly, a relic of bygone times. Uniquely in England, this Nottinghamshire village preserves a truncated but functioning open field landscape. It offers a tiny window onto one of agricultural history's most contentious and enduring enigmas (Beckett 1989). Open field systems and nucleated villages, such as this, once characterised large swathes of medieval England (especially in the midlands) but vanished over the course of the post-medieval period, above all in the parliamentary enclosures of the eighteenth and nineteenth centuries (Rackham 1986, 164–180). The change was both physical and social. The open fields were progressively divided by fences, hedges and ditches, while agricultural activities, previously organised on a communal basis, became more the prerogative of individual landowners. Severalty replaced collectivity.

In its classic midland-type variant, an open field system shares out strips of arable land among its farmers, the strips being (ideally) evenly distributed over two or three large fields which surround a central, nucleated village. The strips are intermingled and lack physical boundaries, such that ploughing, harvesting and other activities must be organised collectively and cooperatively. Crop rotations would likewise be agreed among the villagers, and one field would typically be left fallow and grazed each year. The cooperative bond generated among open field farmers helps to explain the association between open field systems and nucleated villages: such collective agreement and action would be less readily achieved if settlements were dispersed rather than nucleated (Williamson 2003, 155–157).

The origins, operation and fate of this midland-type open field system – this classic 'champion' countryside – have fascinated historians, archaeologists and geographers for well over a century. Strikingly, given the flood of ink spilt on the subject, there is still no agreement over when and how this distinctively tight-knit, communally-organized manner of farming developed in England. In the earlier twentieth century, especially after the publication of Gray's seminal work of 1915, it was conventional to attribute the introduction of champion landscapes to the Anglo-Saxon settlers of the fifth and sixth centuries (Gray 1915, 409–418; Hoskins 1955, 55). Part and parcel of this perceived Germanic package was the heavy mouldboard plough, which churned earth to one side and so created the humped ridge-and-furrow patterns which still corrugate some grassland today: ghosts of arable long since turned over to pasture.

A shift in opinion began around the 1960s, heralded in particular by Thirsk, and was reinforced as it became increasingly clear that there was no archaeological evidence for the traditional model (Thirsk 1964; 1966). Today, nobody is suggesting that the first Anglo-Saxon settlers were pioneers in the empty, wilding wastes of derelict Britannia, carving new fields out of sprawling, untamed woodlands. Documentary evidence shows that champion landscapes were formed and functioning by the thirteenth century, but their origins are not documented at all and remain obscure. Place-name studies, historical geography, settlement archaeology, aerial photography and sophisticated mapping technologies have all been brought to bear on the question in successive projects. For many scholars, the key developments in the history of villages and open fields post-date the period covered by this book. Lewis, Mitchell-Fox and Dyer, for example, propose a crucial 'village moment' between the mid-ninth and early thirteenth centuries (Lewis *et al.* 1997, 227–242).

There is another school of thought, however, which places the main turning point – the germination if not the flowering of open field agriculture – between the seventh and ninth centuries. Focusing principally on Northamptonshire, Hall has argued that open fields were developing by the eighth and ninth centuries, while Brown and Foard posit settlement nucleation around the same time (Hall 1982, 43–55; Brown & Foard 1998). Oosthuizen identifies a long prehistory to the cooperative concepts behind champion farming but, like Hall, argues that the distinctive, extensive systems of medieval England had their more immediate roots in the long eighth century (Oosthuizen 2013a, 89–139). Again, and from a different perspective, Banham has argued that open field farming and its associated ploughing and settlement patterns may have emerged as a response to changing crop preferences in the eighth and ninth centuries (Banham 2010). Moreover, the earliest artefactual evidence for the heavy mouldboard plough in Anglo-Saxon England has now been dated to as early as the seventh century, confounding practically all expectations (Thomas *et al.* 2016).

In sum, it is an increasingly popular theory that open field systems, nucleated villages and heavy ploughs were innovations not of the unstable fifth and sixth centuries, nor of the urbanising tenth to twelfth centuries, but of the economically effervescent seventh to ninth centuries. Clearly, this theory is closely connected to the theme of this book. Does the evidence of fields, farms and ploughs indeed speak of farming transformed in the long eighth century? Did this period really witness the germination of a system whose last flowering lingers still at Laxton?

Fields

Despite their persistent dominance of academic debate, open fields never entirely dominated the medieval English countryside. For centuries, scholars and travellers have recognised a broad distinction between open and hedged

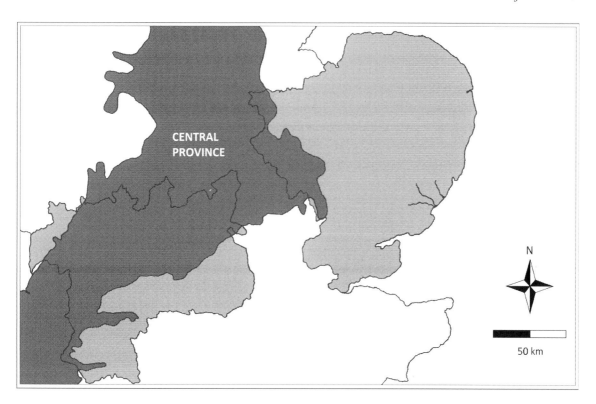

FIGURE 20. Relationship between Roberts and Wrathmell's Central Province and the study regions of this book (using data from Lowerre *et al.* 2011)

landscapes, the 'champion' and the 'bosky', the 'planned' and the 'ancient'. The former in each pair stands for areas where open fields and nucleated villages came to predominate; the latter in each case denotes regions where smaller enclosed plots and dispersed settlements were more common. The most recent mapping of this broad division, based primarily upon nineteenth-century settlement patterns but roughly agreeing with the landscape zones previously mapped by Gray and by Rackham, identifies a Central Province within which open field systems had prevailed in the Middle Ages. The Central Province cuts across this book's study regions (Figure 20), such that, if champion landscapes did evolve between the seventh and ninth centuries, we might reasonably look for evidence of the change here (Gray 1915; Rackham 1986; Roberts & Wrathmell 2000).

Analysis of field systems – their antiquity and development – is a specialist, complex and often legalistic field of study, and original applications are beyond the scope of this book. Most existing scholarship on the question has focused upon Northamptonshire, the real champion heartland, which is situated right in between the two study regions. An arable field of any sort is in a constant state of flux within an overarching cycle of change, and is consequently very hard to date – harder even than a Saxon settlement. Paradoxically, therefore, the basic method applied in Northamptonshire and elsewhere for dating the origins of fields and field systems is largely concerned with settlement evidence.

Having identified the ghost of an open field in surviving earthworks and the configuration of later enclosed plots, the archaeologist must find evidence of a settlement – either by excavation, or from surface scatters of ploughed-up pottery – that lies beneath, and therefore must predate, the field. The latest pottery provides a date for the abandonment of the settlement and, by extension, the earliest possible date for the laying-out of the field. Hence, when Northamptonshire field surveys found Early and Middle Saxon pottery (dated *c*. AD 450–850) in the relict open fields, but none of the ceramics from AD 850 and later, it was inferred that the settlements in question had nucleated and made way for the open fields before AD 850 – most plausibly, during the long eighth century (such work is reviewed by Rippon 2008, 8–10).

There are two problems with this approach. The first, as argued by Oosthuizen, is that we cannot know how much time elapsed between the abandonment of the settlements and the superimposition of the open fields. One may have followed instantaneously upon the other, but that need not have been the case (Oosthuizen 2013b, 96). It is quite possible that an entirely different, intermediary form of land-use – grassy pasture, for instance – succeeded the abandoned settlements but was later obliterated by the laying-out of the open fields at some unknown later date. Still more problematically, early medieval pottery is a notoriously poor dating material. The coarse, friable, handmade wares usually called 'Early/Middle Saxon', especially the undecorated sherds, cannot be dated with any precision. It is not even certain that they had fallen out of use by AD 850 (Mellor 1994, 36–37). Conversely, even the more distinctive, better-dated ceramics of later years were not necessarily ubiquitous throughout their periods of circulation. A humble Saxon farmstead was not obliged to buy in Stamford Ware, say, as soon as it became available in the tenth century, if local coarsewares or wooden vessels were sufficient. Pottery-use may well have been subject to local as well as regional variation. Blinkhorn, for instance, has argued that parts of Oxfordshire could have been essentially aceramic in the eighth century, which would render the period highly elusive in fieldwalking surveys (Blinkhorn in Hardy *et al.* 2003, 172–174).

In sum, therefore, I would more readily accept that rural settlement patterns changed in the long eighth century than that open field systems necessarily emerged at the same time. It may well be that the changes which eventually resulted in the Midland system and its contemporary variants were set in motion around the eighth century, but we are most ill-informed as to how these early steps were taken. One possible prototype, a 'proto-open-field system', has been identified by Oosthuizen in the Bourn Valley in Cambridgeshire (Figure 21), based on her interpretation of a set of very long furlongs spanning four parishes which later operated open field systems (Oosthuizen 2005). She proposes that this proto-open-field system originated between *c*. AD 700 and 917. While the argument is compelling, the proposed chronology is nonetheless still open to question: the *terminus ante quem* of 917 is derived from the presumed but unproven stability of the parish and hundred boundary between the tenth and

FIGURE 21. Location of
key sites mentioned in
the text

nineteenth centuries (Oosthuizen 2005, 178). An arguably more secure *terminus* of 1066 is provided by the Domesday Book and later documents, none of which mentions landholdings spanning four parishes in the way that the long furlongs do.

Thus, we are not yet able to date the origins of open field systems – prototypical or otherwise – to the long eighth century, but it must be acknowledged that *some* transformation of fields and field systems in this period is plausible, given the evidence for changing settlement patterns (discussed further below) and for arable growth which is discussed in Chapter 4. In other words, the evidence and arguments presented in this book all provide a plausible context for the emergence of innovative field systems in the long eighth century, but they do not make an independent case for it.

Meadows

Besides the eponymous arable fields, another important element in the classic open field landscape is the hay meadow. Though not exclusive to open field farming, hay meadows were nonetheless crucial to a champion system which heavily restricted the extent of pasturage in a township. A hay meadow is a piece of managed grassland, closed off for the spring and early summer to allow the grasses to grow, and then mown to provide a store of rich fodder for overwintering livestock (Rackham 1986, 331–334). The resultant ecosystem is dependent on human management and does not survive unchanged if 'left to nature'.

The ecological distinctiveness of hay meadows enables Robinson to interpret changes in the beetle fauna preserved in palaeochannel sediments at Oxey Mead (Oxfordshire), in the Upper Thames valley, in terms of 'a change in the character of the grassland of the site from pasture to hay meadow' (Robinson 2011, 53). Radiocarbon determinations from the sequence suggest that the transformation to hay meadow was underway by the seventh century (Robinson in Hey 2004, 408). While meadow management and hay-making were familiar in Roman Britain, there is no evidence that these practices continued through the fifth and sixth centuries. It may reasonably be posited, therefore, that the evidence at Oxey Mead represents a seventh-century innovation in grassland management.

What remains unclear, however, is how widespread this innovation was at this stage. Documentary references to meadows and hay-making in Ine's law code and Bede's ecclesiastical history, for instance, imply their currency in seventh-century Wessex and eighth-century Northumbria (Banham & Faith 2014, 124). Archaeological evidence, such as it is, tends to be of later date (ninth to eleventh centuries). For instance, Robinson interprets the molluscan evidence from several sites in the Upper Thames valley as an indication that 'hay meadows became extensive in the late Saxon or early medieval period' (Robinson 2011, 53). Pollen and seeds of hay meadow flora likewise occur in deposits of later Saxon date at Market Lavington (Wiltshire) and West Cotton (Northamptonshire), and there is rare charred evidence for hay-making at Abbots Worthy (Hampshire), but this is not closely dated within the Anglo-Saxon period (Wiltshire in Williams & Newman 2006, 136; Campbell 1994, 76–77; Carruthers in Fasham & Whinney 1991, 67–75). Altogether, the evidence is consistent with a seventh-century innovation that continued to spread in the centuries thereafter, albeit slowly and unevenly (Rackham 1986, 334–335).

Ploughs

In the summer of 2010, excavations at Lyminge (Kent) unearthed a unique find in early Anglo-Saxon archaeology: an iron coulter, the vertical slicing element from a heavy plough (Figure 22). Discovered at the base of a *Grubenhaus* and firmly dated to the earlier seventh century, the coulter constitutes the earliest evidence for a heavy plough in Anglo-Saxon England (Pitts 2011; Thomas *et al.* 2016). The fact that the coulter had been carefully – perhaps ritually – deposited, rather than recycled, has not only resulted in a discovery momentous enough to be reported in national news. It also suggests that the object held some special significance for the inhabitants of this Kentish royal settlement in

FIGURE 22. Simplified comparison of plough types (after Bowen 1961, 8, fig. 1)

1) ard

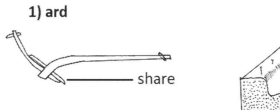
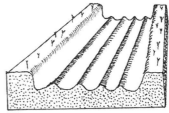

2) heavy mouldboard plough

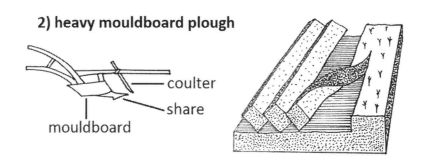

seventh-century England: functional significance, as the coulter shows signs of physical wear, and cultural significance, redolent of contemporary customs in the Frankish realm (Thomas *et al.* 2016, 753–755).

It is a rare enough event to recover any agricultural artefact from an Anglo-Saxon settlement, let alone such a significant coulter. The surviving corpus of Romano-British and earlier farming tools appears curiously profligate by comparison (Rees 1979). The majority of early medieval implements will have been made of wood or iron, neither of which survives well archaeologically: the former usually decaying, the latter being vulnerable to corrosion or else recycling as a valuable resource in antiquity. Surviving artefacts of agrarian function, ferrous or otherwise, are therefore very scarce within the study regions, as elsewhere. Bells, such as those found whole or in part at Shakenoak (Oxfordshire) and Dorney (Buckinghamshire), may originally have been attached to livestock, but a pastoral function is entirely conjectural and would, in any case, only indicate the predictable fact that some livestock had been free-ranging (see, for example, Brodribb *et al.* 2005, 238). Other finds include a range of cutting implements: shears, sickles, pruning and reaping hooks, few of which can be securely identified or dated. For many of them, an exclusively agricultural function need not be assumed. Shears such as those found at Yarnton (Oxfordshire), for instance, although they may have had broadly agrarian uses such as for wool-clipping or straw-cutting, could well have served other functions, such as in textile working (Hey 2004, 82).

In short, finds of Anglo-Saxon farming tools are too few, fragmentary, or undiagnostic to admit useful comparative study. This evidence seems consistent with a lack of anything resembling an industry or market for tools – a scenario in which local, occasional production likely prevailed, and in which widespread innovation was therefore slow to take hold amongst disparate craftsmen.

By contrast, the Lyminge coulter is important precisely *because* it is an exceptional find, so far unparalleled in fifth- to ninth-century England, shedding a rare light on the vexed issue of Anglo-Saxon ploughing. As with many of the topics considered in this book – including, especially, the field systems discussed above – past scholarship has often taken a teleological perspective on the development of ploughing. According to the traditional narrative, Anglo-Saxon settlers introduced the heavy mouldboard plough to Britain in the fifth and sixth centuries, replacing the lighter ploughs of their predecessors, and thus setting the pattern for medieval and later English tillage (Hoskins 1955, 55). Although this model has long since been disputed, the question which it aimed to answer has remained current and unresolved: how, when and why did the heavy mouldboard plough enter English agricultural history?

This question has significance beyond the arcane, aratral interests of tillage enthusiasts. At its heart is a simplified dichotomy between two types of ploughing instrument and method. The so-called heavy plough is distinct from the lighter ard – or 'scratch plough' – in important functional ways (Fowler 2002, 182–196). The ard is equipped solely with a share (OE *scear*), to shear grooves through the soil. To the heavy plough is added a coulter (OE *culter*) to cut vertical slices through the soil; and a mouldboard, a curved plate which inverts the sod, and so not only breaks up the tilth but also aerates it and upsets weed growth (Figure 22).

There is a danger of reducing the debate to a simple question of whether and when the Anglo-Saxons switched from using the ard to using the mouldboard plough, as if the *gens Anglorum* had collectively upgraded their tillage equipment. In fact, as Banham and Faith argue, the ard is likely to have been only gradually and partially replaced by the heavy plough in this period, since not all farmers will have had sufficient resources, appropriate terrain, or a strong enough inclination to adopt the heavy plough (Banham & Faith 2014, 44–50). While the mouldboard plough would have afforded better action against weeds and a more thorough distribution of nutrients in the tilth – and would also have been essential for the extensive cultivation of heavy clay soils – one cannot presume that supposed economic rationality was always sufficient recommendation for overturning ancient tillage customs (Williamson 2003, 120–122). The much greater investment in wood, iron, labour and oxen required for the heavy plough may have restricted this innovation to the wealthier, or more communally-minded, farmsteads.

A form of heavy plough, if not a true mouldboard plough, seems to have existed in Roman Britain (Booth *et al.* 2007, 288), but there is no direct evidence for its contiguous survival into Anglo-Saxon England. The seventh-century Lyminge coulter represents a specific type known as a swivel plough, familiar from contemporary Frankish contexts and plausibly interpreted in the context of continental influence upon Anglo-Saxon England – Kent in particular – at this time (Thomas *et al.* 2016, 752–753). By the later Anglo-Saxon period, evidence is more forthcoming. More shares and coulters are known from the ninth century onwards, such as at the high-status sites of Flixborough (Lincolnshire) and Bishopstone (East Sussex) (Ottaway in Evans & Loveluck 2009, 245; Thomas 2010, 130). The tenth-century Exeter Book contains a riddle clearly describing the action of a mouldboard plough, and some eleventh-century illustrations imply the artists' familiarity with the same technology, at least in the élite milieu of manuscript production (see the evidence assembled by Hill 2000, 11–13; Banham & Faith 2014, 46–50). At Drayton (Oxfordshire), a buried soil was found to have a cross-section consistent with the inverting motion of a mouldboard plough; extrapolation from archaeomagnetic readings suggested a date between the ninth and eleventh centuries for the ploughing (Barclay *et al.* 2003, 115–116). This evidence seems to occur only as an isolated stretch, and has therefore been interpreted as 'a *single* episode to break up an area of compaction' (Booth *et al.* 2007, 333, my emphasis). This 'single-use' interpretation does not quite allow us to infer that heavy ploughing had become a regular, well-established practice in later Saxon Oxfordshire. And it is perhaps significant that the parish of Drayton, like Lyminge, is likely to have been the location of a royal central place in the seventh century (Brennan & Hamerow 2015).

The limited evidence for heavy ploughs in Anglo-Saxon England therefore begins in the lordly context of seventh-century Kentish kings and, while it had clearly spread geographically by the ninth, tenth and eleventh centuries, there is no sign that the technology had spread any further down the social scale with the passage of time. Throughout this period, considering the evidence detailed above, the mouldboard plough may well have remained an exclusively élite innovation.

Farms

The previous chapter showed how unevenly distributed are the excavated Anglo-Saxon settlement sites included in this study, and remarked upon how their skewed distribution may be due in part to the exigencies of archaeological survival and discovery. There are far denser concentrations of sites recorded around the Upper Thames and Great Ouse and their tributaries than anywhere else. The East Anglian Plain and various chalk and limestone uplands are poorly represented by comparison (see Chapter 1). Some factors influencing this distribution are essentially arbitrary, such as the variable effects of funding, scheduling and urban development upon the volume and accessibility of data. The clusters of evidence at Ipswich, Milton Keynes, and the Walton area of Aylesbury all reflect concentrations of urban development-led excavations, for instance. Large-scale modern gravel extraction has also led to the discovery and excavation of major sites such as Mucking (Essex) and Yarnton (Oxfordshire), and may to some extent be responsible for the marked distributional bias towards river valley locations in general and the gravel terraces in particular.

Gravelly soils are also conducive to both cropmark identification and the preservation of those organic remains which have been targeted especially by this study (Robinson 1992, 48; Booth *et al.* 2007, 1; Hamerow 2012, 3). On the other hand, such locations may well have been genuinely attractive to Saxon farmers, offering ready access to water, riverine transport, light, easily worked soils, and the potential for lush, alluvially enriched grassland.

All of this matters because it is the key to understanding a long-perceived phenomenon which looms large in studies of Anglo-Saxon rural settlements – or farms, as they may reasonably be called in this instance. It is variously known as the 'Middle Saxon shift' or 'shuffle', and deserves particular attention here. As early as 1963 it was observed that the archaeologically detected settlement patterns of the fifth and sixth centuries appeared different from those recorded in the Domesday survey of 1086. Clearly some dislocation had occurred between the fifth and eleventh centuries (Hunter Blair 1963, 269–270). In 1981, Arnold and Wardle argued that such a widespread dislocation occurred in the later seventh and early eighth centuries, in the first explicit articulation of the 'Middle Saxon shuffle' model (Arnold & Wardle 1981). Artefactual chronologies from various excavated farms suggested that those established in the fifth and sixth centuries were abandoned, and replaced in new locations, around the later seventh century. Hodges reaffirmed the model, and emphasised the topographical dimension, which is of agricultural significance: the new farms founded from the late seventh century onwards occupied heavier, more fertile soils, as opposed to the 'inferior' soils occupied by the abandoned earlier settlements such as West Stow, Suffolk (Hodges 1989, 62).

The model has since been subject to modification and critique, partly from the recognition that the establishment and abandonment of early medieval farms are difficult to date with any great precision; and partly also from Hamerow's argument that some earlier settlements, such as Mucking, Essex, shifted or drifted throughout their periods of occupation (Hamerow 1991; 2012, 67–70). The seventh century might then have witnessed not so much a radical dislocation, as the emergence of greater stability in settlement patterns, perhaps alongside an expansion of settlement onto heavier soils. The earlier 'wandering' of rural settlements may have been a necessary response to the paucity of their soils and low intensity of their farming strategies: soils would be periodically abandoned once exhausted or weed-infested. By contrast, settlement stability could be associated with longer-term exploitation of more fertile soils in a more labour-intensive regime, the fields being kept in good heart through techniques such as manuring, fallowing and crop-rotation (Fowler 2002, 208–212; Williamson 2003, 121).

A stabilisation of rural settlement patterns would have been directly conducive to agricultural intensification, since labour, ploughs and other resources may be more effectively deployed if stable and centralised (Williamson 2003, 67–68, 157). Indeed, settlement patterns could have especially profound implications for our understanding of tillage. Williamson, for example, has directly adduced the changed settlement patterns of this period as evidence for the antiquity of heavy ploughing:

> 'there can be little doubt that they [mouldboard ploughs] had come into widespread use in middle Saxon times simply because, by the eighth century, settlement had expanded once more onto the kinds of heavy soil which could not be easily cultivated without them'. (Williamson 2003, 120)

Besides the evidence from excavated farms, fieldwalking surveys of surface scatters of potsherds – already mentioned above in the discussion of field systems – have also contributed to this debate about settlements. The results of the Sandlings survey in south-east Suffolk, for instance, appear to show the reoccupation of heavier soils from the seventh and eighth centuries onwards, following a post-Roman retrenchment to the lighter, sandier ground (Figure 23; Newman 1992).

The survey evidence from Northamptonshire, as discussed by Brown and Foard, offers a subtly different picture. Here, the river valleys are described as offering 'more favourable soils', preferred throughout the fifth to ninth centuries, following 'an almost complete loss of settlement from the marginal land, that is, from the claylands on the watersheds between the main river valleys' by the early fifth century (Brown & Foard 1998, 73–75). Rather than *expansion* of settlement patterns, they argued that *nucleation* was the defining trend of the eighth and earlier ninth centuries – nucleation which entailed the abandonment of earlier dispersed farms (over whose remains the open fields eventually spread) in favour of centralised settlements, which were the precursors of medieval manors and villages. In this way, Brown and Foard

FIGURE 23. Changing settlement patterns over time in the Sandlings area of Suffolk (after Newman 1992, 31–35, figs 6 & 8)

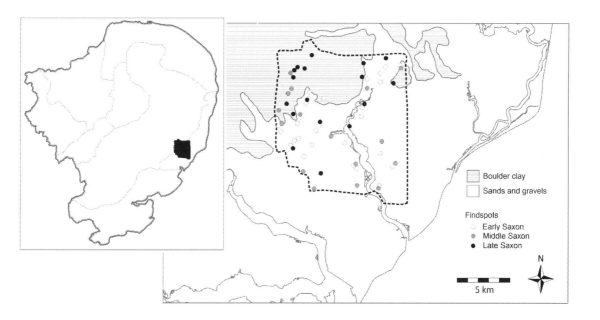

place the first stages of 'manorialisation', whereby peasant cultivators lost their freedom to increasingly powerful landlords, in the long eighth century (Brown & Foard 1998, 91).

The layouts of excavated settlements lend some support to the notion that settlement nucleation began between the seventh and ninth centuries. Farms of this period, such as that excavated at Pennyland (Buckinghamshire), do not particularly resemble nucleated villages in the champion sense; they appear to have very few dwellings, for one thing (*pace* Williamson 2003, 67); but they do represent a clear departure from the earlier Saxon norm. Citing many examples, Reynolds has observed that axial alignment of buildings and the use of rectilinear boundary features – 'suggestive of imposed spatial regulation' – appear to be an innovation in rural settlements from the late sixth or early seventh century onwards, in stark contrast with their more mobile, dispersed, unenclosed and relatively disorderly predecessors (Reynolds 2003, 110–119).

All things considered, therefore, studies of Anglo-Saxon rural settlement forms and patterns over the past three decades have provided a compelling, circumstantial argument for agricultural change in the seventh to ninth centuries. The question remains, however, as to what kinds of agricultural change are implied by the changing forms and distributions of Anglo-Saxon farmsteads? It is important here to retain the connection between the two aspects of this question. The distribution patterns of farms, and their internal layouts (that is, the morphology of their excavated plans), are likely to be related phenomena, since both will have depended upon the kinds of activity which farmers wanted or expected to undertake.

Is there any evidence, in the project dataset, of a return to the clays between the seventh and ninth centuries, as posited in the 'Middle Saxon shift' model? Chapter 1 described the environmental contexts of the sites recorded in the dataset, and found relatively few sites to occupy thoroughly clayey ground, even in some areas characterised by clays. For instance, in contrast to the abundant evidence from the gravel terraces of the Upper Thames valley, the intervening rises of Jurassic clay are virtually barren of data, even into the long eighth century. Conversely, in the vicinity of Milton Keynes, clay soils seem to have been occupied to some extent throughout the fifth to ninth centuries, with no obvious post-Roman hiatus (see, for example, Bancroft, Wolverton Turn, or Westbury-by-Shenley). In only two areas does it really appear that clay terrain was largely abandoned between the Roman period and the seventh or eighth century: Ely, in the peat fens, where the West Fen Road occupation sequence begins around the eighth century; and north-west Essex, in the southern portion of the East Anglian Plain, where evidence appears at Stansted and Wicken Bonhunt from the seventh century onwards. Admittedly this may not be a truly representative survey of the regions, since I have not aimed to record all traces of Anglo-Saxon settlement, but it does begin to illustrate the complexity of known settlement patterns.

An additional complication, seldom discussed but worth considering, is the difference between occupation and usage. If clay terrain was never settled, it may nonetheless have been used for grazing by surrounding farms. Seasonal grazing within a tradition of transhumant pastoralism – entailing the periodic movement of livestock to pastures distant from their home settlements – would leave little trace in the archaeological record. Temporary occupation by shepherds, for instance, may have been ephemeral in the extreme. With this possibility in mind, we should return to consider the general patterns of settlement distribution in the study regions, as described in Chapter 1. It was stated there that most sites lie at, or close to, the junction of two or more types of terrain. This is a crucial point. For our purposes, the soils immediately beneath and surrounding an excavated farm are less important than the full range of those within its wider hinterland, i.e. those soils which are likely to have constituted its farmland. For most sites in the project dataset, this hinterland most likely took in sands, gravels and clays, plus chalk and limestone uplands for a few sites. The hinterland of a typical farm may have had sands and gravels on the river valley terraces – sheltered, well-drained, but also well-watered, and therefore favourable for domestic settlement – and intervening clay ridges as watersheds. This is exactly the situation described by Williamson, focusing particularly upon Norfolk, who deduces from documentary and place-name evidence that the raised

FIGURE 24. Map of sites with evidence for raised causeways in the study regions

interfluves (watersheds), clayey or otherwise, tended to support woodland, pasture, or indeed wood-pasture in the early medieval period (Williamson 2003, 37–43).

Direct evidence of transhumance in the fifth to ninth centuries is literally thin on the ground: the practice is not such as to leave much of an archaeological footprint. One rare category of evidence is, however, at least consistent with models of transhumant pastoralism: the raised causeway or river crossing (Figure 24). One possible example has been excavated at Scole, where two parallel alignments of oak piles were found along the edge of a palaeochannel in the Waveney valley (Suffolk). Selected piles were radiocarbon-dated, and returned determinations spanning the fourth to seventh centuries (Ashwin & Tester 2014, 199). The causeway had been erected by the sixth century at the latest. Another causeway, the Strood on the Essex coast, must have been built or repaired in the seventh and eighth centuries: dendrochronological analysis of its oak piles returned felling dates *c.* AD 684–702 (Crummy *et al.* 1982; cf. Brunning 2010 for a comparable structure of similar date, beyond the study regions in the Somerset Levels). Finally, successive excavations in the St Aldate's area of Oxford have produced evidence of artificial crossings between alluvial islands on the floodplain. A cobbled surface at the BT Tunnel site could represent a causeway-cum-ford, possibly in use by the late sixth or early seventh century, since a plank in the overlying silt was felled *c.* 577–619. Cattle hoof-prints preserved at the nearby Police Station site, at a comparable place in the alluvial sequence, hint more directly at livestock-movement among the floodplain islands (Dodd 2003, 12–13). Later evidence includes a putative bridge trestle at the BT Tunnel site (cal. AD 660–900), wattle revetments at the Trill Mill Stream site (cal. AD 660–990), and a complex sequence at 79–80 St Aldate's, including further revetments and a clay bank which may represent a causeway, thermoluminescence-dated to AD 631–779 (Dodd 2003, 15–16; Durham 1977, 176–179).

Hence, there is a limited and debateable set of causeway sites which may reflect transhumant practices spanning the entire fifth to ninth century period. This practice, at least, does not seem to have been an innovation of the long eighth century, and there is no particular reason why it should have been. It has already been noted that there is ecological evidence for grassland pasture in the Upper Thames valley before the seventh century, and circumstantial evidence (i.e. the dearth of settlement/pottery) consistent with the existence of wood-pasture on the interfluvial clays, at least until the seventh century. However, the evidence discussed above also suggests that some riverine pastures were, from the seventh century onwards, managed as hay meadows; and the survey evidence from the Sandlings and elsewhere could reflect the expansion of arable farms onto clayey watersheds at around the same time. These two phenomena – increased cultivation on clay soils and the introduction of hay meadows – are likely to have been related. The creation of hay meadows implies a need for stores of winter fodder; the loss of wood-pasture to arable on the clay interfluves would imply a reduction of winter grazing, since woodland provides grazing matter into the winter, after grassland has died back (Williamson 2003, 163–166).

Arable expansion, in this model, placed greater strictures on transhumant pastoralism by encroaching upon winter pasture, and so encouraged hay meadow management to provide fodder for livestock now being overwintered closer to their parent farmsteads. It is this part of the model, the overwintering of livestock within rural settlements rather than on winter wood-pasture, that brings us back to the forms of farms, and in particular the rectilinear boundary features described by Reynolds as 'suggestive of imposed spatial regulation' (Reynolds 2003, 119).

As Reynolds observes, and as Hamerow has elaborated, there is a phenomenon of 'rectilinear settlements', none of which can be dated any earlier than the seventh century. These are characterised by 'rectilinear paddocks, trackways, plots, and field boundaries', as opposed to enclosures associated with particular buildings (Hamerow 2012, 72–73). Sometimes these rectilinear features were imposed upon earlier Saxon settlement sites, and sometimes the entire settlement was laid out as such from the outset; but a starting point for the whole rectilinear phenomenon belongs, on current evidence, in the seventh century (Hamerow 2012, 73–83). While these and other changes are often seen through the teleological lens of nucleation, and the ultimate emergence of the classic champion village, there has been less debate over the *functions* of the rectilinear features, and there is certainly no consensus. Wright, for instance, considering the evidence from Ely and Fordham (Cambridgeshire) and Foxley (Wiltshire), interprets these sites as monastic dependencies, and argues that the rectilinear divisions were an articulation of the parent minster's lordly power over its dependent peasants: 'Property and household plots were defined with great rigour and with greater permanence so that the social space of individuals could be used to reflect their place in the social hierarchy' (Wright 2015, 26). Hamerow, by contrast, offers a more pastoral interpretation, suggesting that arable expansion, loss of pasture, management of hay meadows, and the need rapidly to assemble teams of plough oxen led to animals being kept closer to settlements and therefore requiring closer spatial control. She further proposes that a concentration of 'rectilinear settlements' around the fen-edge could reflect a particular lack of winter-grazing in this wet environment combined with the need for rapid ploughing on heavy clay terrains (Hamerow 2012, 88–90). Wright acknowledges this argument but contends that monastic 'home farm' dependencies at Ely, Fordham and Foxley represent a particular type of site with particular social factors behind their development (Wright 2015, 39). Given the evidence and ideas discussed so far in this chapter, however, I believe there is a strong case for adopting a pastoral interpretation for many if not all of these sites, monastic or otherwise.

I have attempted to identify all excavated settlements within the study regions which have what I will term paddock and droveway complexes. These are rectilinear ditch systems which, despite often being very patchy in plan, exhibit one or both of these patterns: extended parallel linear ditches (the droveways) and rectilinear ditched enclosures with few or no structures inside them (the paddocks). Besides an absence of domestic occupation evidence, it is difficult to find evidence in direct support of a pastoral interpretation of an empty enclosure; livestock, regrettably, do not use pottery. However, some morphological clues are proposed by Pryor in his discussion of Bronze Age stockyards in the fens. These clues include the use of corner entranceways and other funnelling devices to help channel livestock (especially sheep) which are otherwise difficult to herd into and out of enclosures (Pryor 1996). On these grounds, I have identified several relevant sites in the study regions (Figure 25). While droveway or paddock interpretations are not incontrovertible for these sites, I consider them to be at least consistent with the known morphologies of the excavated ditch systems. The selection here is by no means a comprehensive inventory of sites in the study regions with rectilinear ditched features. However, those which have been omitted from the list are either too small, poorly dated, or morphologically indistinct to be plausibly interpreted in this way (such as at Market Lavington, Wiltshire); or else they are of such different size and shape as to be considered a different phenomenon altogether, needing alternative interpretations (such as the vast enclosure at Wolverton Turn, Buckinghamshire).

FIGURE 25. Map of sites with paddock and droveway complexes in the study regions

Plans of the rectilinear ditch systems preserved at these sites are reproduced in Figures 26 to 32. Many of these plans are difficult to read and interpret because of lacunae and complex intercutting, overlapping and re-cutting of features. Few, if any, appear to represent a single, unchanged, planned layout, which perhaps reflects the longevity of this phenomenon: these ditch systems were clearly around for long enough to warrant renovation. However, there is a persistent pattern of droveways running parallel to paddocks, which in some cases – such as at Gamlingay and Godmanchester – do seem to have the corner entrances described by Pryor (Figure 33). Indeed, at Godmanchester there are hints of more complex divisions, conceivably designed for the sorting of livestock in the manner suggested by Pryor for the Fengate sites (Figure 34; Pryor 1996).

Most complicated of all, however, are the vast ditched systems excavated at West Fen Road, Ely, which sheer proximity suggests were associated with the minster on that fenland isle (Mortimer *et al.* 2005; Mudd & Webster 2011; Wright 2015, 35–39). Some of the enclosures may represent domestic occupation plots, perhaps the hierarchical settlement divisions envisaged by Wright. But taken as a whole, the complex bears a significant resemblance to the Bronze Age stockyards in Pryor's paper, with multiple parallel droveways, various sizes and configurations of paddocks or yards, corner entrances, and plausible funnels or 'crushes' to aid herding and sorting. Following Pryor's example, we might estimate that flocks running to hundreds if not thousands of sheep once crowded through these extensive systems (Pryor 1996, 317). As with all the other examples discussed here, however, we cannot necessarily judge which enclosures were used for cattle, sheep, pigs or even horses: the first three species are, after all, ubiquitous among the animal bones from all sites in the project dataset (see Chapter 3).

The distribution of these supposedly pastoral complexes is also instructive. Most sites are indeed focused around the Fens, as Hamerow observed, but examples also occur in the Milton Keynes area of the Midland Clays (at Bletchley and Pennyland), in the Upper Thames valley (at Yarnton and Lechlade), and to the south at Wantage (Oxfordshire). Those sites which, unlike Ely or Wantage for example, do not lie immediately adjacent to clayey terrains, are nonetheless well-placed to have clays in their wider hinterland: the Oxford Clays of the Upper Thames valley, or the Midland and fen-edge clays in the basin of the Wash and Great Ouse (Figure 25).

Recalling the points raised earlier in this chapter, therefore, it is possible to offer a consistent pastoral interpretation for all of these sites. As arable land expanded in the seventh and eighth centuries, and marginal, clayey winter pastures were encroached upon, farmers began to develop hay meadows and elaborate rectilinear stockyard systems with the combined aims of (i) overwintering livestock locally, (ii) enabling the rapid mobilisation of plough oxen on poorly-drained farmland, and (iii) protecting crops from the ranging mouths of hungry ungulates. In this model, rectilinear settlements were designed primarily to keep animals, not people, in their proper place.

Conclusions

This chapter has marshalled a range of evidence – mostly archaeological, some historical, and a fair amount entirely circumstantial – to investigate how fields, ploughs and farms developed between the fifth and ninth centuries. Most of this evidence, and its interpretations, remain equivocal, especially the material pertaining to the much-debated Midland system and its open fields, nucleated villages and heavy ploughs. This book does not claim to offer any original take on the long-running open field debate, save to suggest that more evidence is needed before a more conclusive story can be told. However, another narrative has begun to emerge: potentially, but not necessarily, independent of the open field saga.

A tiny, diverse collection of sources suggests that the heavy mouldboard plough had arrived in Anglo-Saxon England by the seventh century, in a Kentish royal context, and then spread across the country (but perhaps not all the way down the social spectrum) to become familiar in élite Anglo-Saxon circles by the tenth and eleventh centuries. Such ploughs offered improvements in crop husbandry in return for heavy investments in labour and materials. They were especially valuable, arguably essential, on clay soils. This technology may therefore have facilitated an expansion of arable farming (and, with it, long-term settlement) onto clays from the seventh century onwards, as evidenced in some field surveys. I have speculated that such arable expansion onto clay watersheds came at the expense of lost winter pasture or wood-pasture on these hitherto marginal lands, and therefore began to constrict patterns of transhumant pastoralism. In response, again from the seventh century onwards, farmers began to manage and mow hay meadows in order to provide fodder for livestock overwintered closer to the farm. This in turn necessitated the greater spatial control of livestock in order to protect crops, and indeed thatched buildings, from the wandering beasts. Hence there emerged at this time the droveway and paddock complexes, which furthermore will have facilitated the rapid deployment of oxen to pull the heavy plough. So far, within the study regions and the seventh- to ninth-century period, the most persuasive evidence for this entire model comes from the Upper Thames valley and the Midland Clays, taking in the basin of the Wash and Great Ouse.

I argue that this is an internally consistent interpretation of the available evidence, but I also concede that it is largely circumstantial, offering little or no independent evidence for the growth of arable or for increased control over livestock. In the next three chapters, therefore, I explore animal husbandry and arable farming in closer detail, drawing especially upon animal bones and plant remains. It is time to put some flesh on the bones of this story.

FIGURE 26. Plans of paddock and droveway complexes in the study regions at (A) Bletchley (after Hancock 2010, 10, fig. 4); (B) Orton Waterville (after Wright 2004, 25, fig. 2); (C) Downham Market Bypass (after Percival 2001, fig. 3); (D) Wantage (after Holbrook & Thomas 1996, 126, fig. 9); (E) London Road, Downham Market (after Trimble 2001, fig. 9); (F) Harston Mill (after O'Brien 2016)

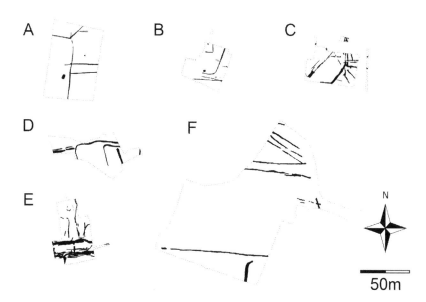

FIGURE 27. Plan of paddock and droveway complex at Pennyland (after Williams 1993, 94, fig. 52)

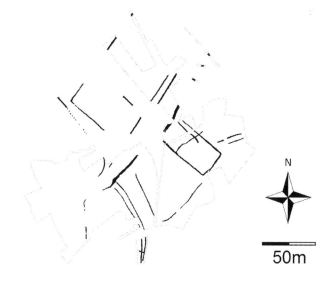

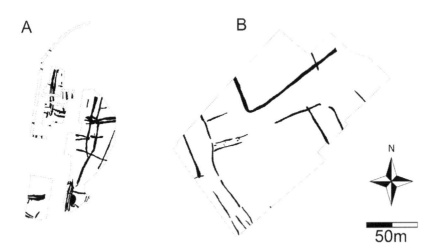

FIGURE 28. Plans of paddock and droveway complexes in the study regions at (A) Lakenheath (after Caruth 2006, 73, fig. 43); (B) Godmanchester (after Gibson 2003, 204, fig. 42)

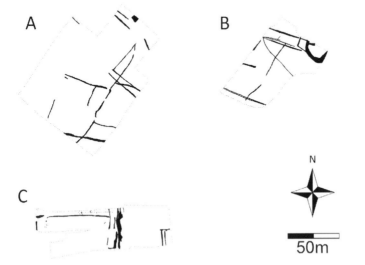

FIGURE 29. Plans of paddock and droveway complexes in the study regions at (A) Gamlingay (after Murray 2005, 182, fig. 10); (B) Lechlade (after Bateman *et al.* 2003, 41, fig. 12); (C) Fordham (after Patrick & Rátkai 2011, 43, fig.3.2)

FIGURE 30. Plans of
paddock and droveway
complexes at Cottenham,
two successive phases
(after Mortimer 2000,
9–10, figs 7 & 9)

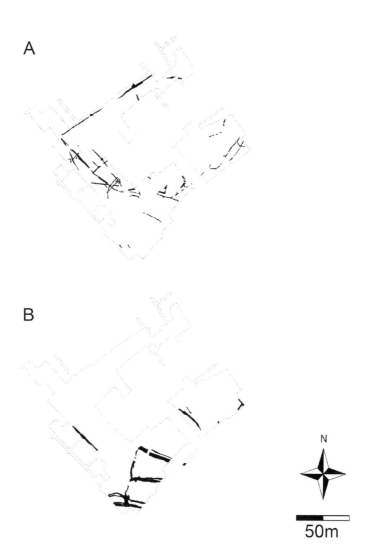

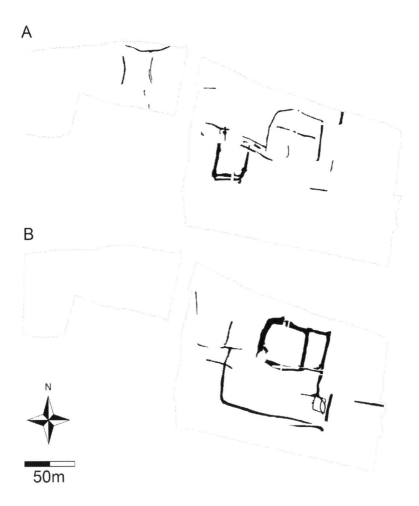

FIGURE 31. Plans of paddock and droveway complexes at Yarnton: (A) Phase 2; (B) Phase 3 (after Hey 2004, 20, fig. 1.5)

N

50m

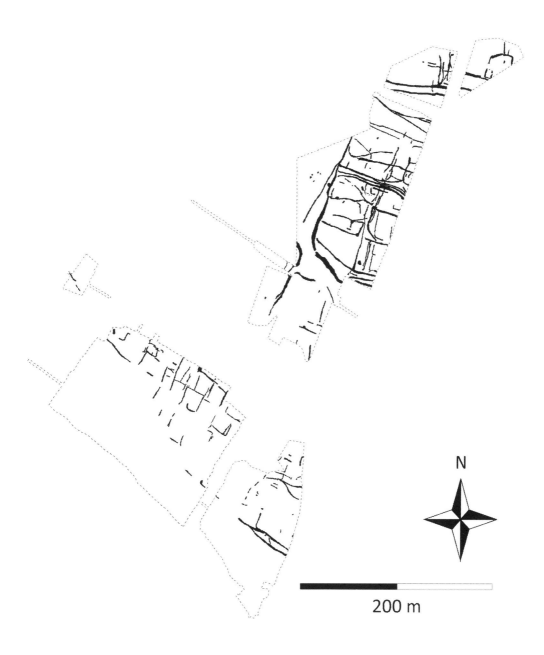

FIGURE 32. Plan of paddock and droveway complex at West Fen Road, Ely (after Mortimer 2005, 145, fig. 7.1: Ashwell Site to the south, Consortium site to the north)

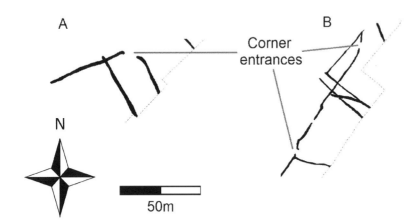

A

B

Corner
entrances

N

50m

FIGURE 33. Details of
(A) Godmanchester and
(B) Gamlingay plans,
showing corner entrances
(after Gibson 2003, 204,
fig. 42 and Murray 2005,
182, fig. 10)

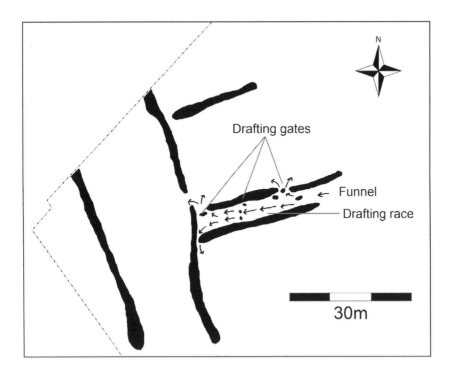

N

Drafting gates

Funnel

Drafting race

30m

FIGURE 34. Detail
of Godmanchester
plan showing possible
livestock-sorting
arrangement (after
Gibson 2003, 204,
fig. 42)

CHAPTER 3

Beast and bone

It is a truth universally acknowledged, that a Saxon farmer in possession of an empty field must be in want of a sheep. And if the acknowledgement is not quite universal, and the truth not quite proven, it is a popular enough theory. In scholarly literature and anecdotally, the story of early medieval English farming is very largely a story about sheep. When discussing paddocks and droveways in the previous chapter, I drifted almost automatically towards ovine interpretations (see Chapter 2). On purely theoretical grounds, this view is entirely reasonable. For the Anglo-Saxon peasant, operating at (or little above) subsistence level, sheep seem a credibly utilitarian choice of livestock, providing milk, meat, hide, horn, bone, yarn and manure, and thriving inexpensively on rougher pasture (Banham & Faith 2014, 92). If only one could shelter beneath it too, *Ovis aries* would practically offer subsistence on four legs. Conversely, for the merchants of the age, or for gift-exchanging lords, sheep's wool may have been the crucial good that crossed the Channel, foreshadowing the famously prosperous wool trade of the thirteenth and fourteenth centuries. It is often noted that Mercian cloaks, presumed woollen, were a significant enough export to feature in a letter from the Frankish king Charlemagne to the Mercian king Offa in AD 796: 'our people make a demand about the size of the cloaks, that you may order them to be such as used to come to us in former times' (EHD no. 197).

Sheep are sometimes deemed such a likely source of wealth in the long eighth century that their importance is adduced with little or no reference to their physical remains. Metcalf, for instance, postulates a trade in Cotswold wool to explain a concentration of eighth-century coin-finds near the Upper Thames (Metcalf 1977, 99; 2003, 43–45). Likewise Ulmschneider, assessing the wealth of seventh- to ninth-century coinage and metalwork in Lincolnshire, Hampshire and the Isle of Wight, suggests ovine origins to this evident prosperity (Ulmschneider 2000, 94). I argued in the last chapter that paddocks, droveways and hay meadows were a marked innovation in agricultural landscapes in the seventh to ninth centuries. Populated with sheep, their theoretical significance could become seductively clear: they would be productive investments in a nascent market economy, ultimately delivering up wool to fetch a fine price at the docks of Quentovic or Dorestad across the water.

So far we have spun a more or less feasible yarn with little supporting evidence. However, this 'woolly hypothesis' is not entirely conjectural. Blair has argued for the importance of wool production in Anglo-Saxon Oxfordshire, citing not only the many weaving combs, pin beaters, and loom weights discovered at excavated settlements but also the evidence of sheep bones, and their tendency to dominate Anglo-Saxon faunal assemblages. Indeed, several zooarchaeologists, studying the faunal remains discovered at Anglo-Saxon settlements, have found supporting evidence for the importance of sheep and wool, both at individual sites and in synthetic studies, though with differing chronologies and interpretations. 'The use of cattle as traction animals in the Middle Saxon period,' writes Crabtree, 'is accompanied by the increasing importance of sheep at many Anglo-Saxon sites' (Crabtree 2012, 58). Payne, by contrast, suggests a later ascendancy:

> 'Sheep bones tend to be commoner in late Saxon sites … and documentary references suggest that wool production was of increasing importance in the later Saxon period, both trends foreshadowing the economic importance of the later medieval wool and textile industries' (Payne 1999, 39).

Clutton-Brock simply proposed that 'sheep farming probably grew into Britain's major industry during the Anglo-Saxon period' (Clutton-Brock 1976, 380), while Holmes has identified 'a predominance of sheep on the majority of rural and ecclesiastical sites in early-late [Saxon] phases' (Holmes 2014, 40). For O'Connor, meanwhile, the trend is regional: 'it seems … likely that sheep were particularly abundant in early to mid Saxon central East Anglia' (O'Connor 2011, 367).

There are, however, some dissenting voices. Sykes contends that cattle remained the most important species of livestock until after the Norman Conquest, with sheep in second place except at a few 'precocious' sites in Oxfordshire and Buckinghamshire (Sykes in Preston 2007, 107–108). Tipper also questions the importance of sheep husbandry and wool production, looking specifically at the evidence from West Stow (Suffolk). A supposed rise in the proportion of sheep bones at that site is, he argues, only slight and temporary, and the overall quantities too meagre to indicate surplus production on any great scale (Tipper 2004, 177–182).

This presents us with a conundrum. When it comes to elucidating animal husbandry in antiquity, the bones of the beasts bear more direct witness that anything else, but their testimony is unclear. More precisely, it is unclear as to how we should interpret the evidence of the bones, the more so as statements drift further from the original data. Sheep bones are certainly not rare in Anglo-Saxon archaeology, but it is no simple matter to argue from a few cracked vertebrae to a wool boom. It is all too easy to paraphrase 'an increase in the proportion of sheep bones' as 'more sheep', and thus obscure the first observation with an abrupt, unqualified interpretation.

Suppose we were discussing changes in the livestock populations of a modern farm. An increase in the proportion of sheep would not necessarily mean 'more sheep', i.e. an increase in the absolute numbers of rams and ewes. Total livestock numbers could actually have declined, but sheep less so than cattle or pigs, meaning a higher proportion of sheep but fewer sheep overall. But in zooarchaeology, we are not even discussing biological populations. Rather, we are discussing skeletal remains, which are the best available proxy for the original populations. How good a proxy the bones may be is open to debate, and depends partly upon *which* original populations we are trying to study. We may wish to study the entire population of livestock bred and raised at a given farm, in each phase of its occupation. Yet the bones can only ever represent the livestock *consumed* at a site. Consumption could include bone-working as well as meat-eating, such that breeding, raising, slaughtering, butchering, and two distinct stages of consumption could each have taken place at a different settlement, with little or nothing left by way of an audit trail for the archaeologist to follow. In any case, we are unlikely to discover every animal bone that was ever deposited at a site, or even a representative sample. Some will have deteriorated beyond recovery or recognition, others will lie outside the excavated area, and still others may be of uncertain date or species.

Put concisely, the relationship between original livestock populations and fragmentary archaeological deadstock populations may appear hopelessly jumbled and obscure. Except with those very few faunal assemblages which are extremely well preserved, and which have been subject to the most detailed forensic analyses, we cannot begin to approach ancient livestock populations in a biologically meaningful way. I mean not to give an unremittingly bleak verdict on the archaeological value of animal bones in studying ancient pastoralism, but rather to focus on what they may or may not truly signify in terms of animal husbandry. This book is not a strictly zooarchaeological study, and attempts no original, first-hand analyses of animal bone assemblages. It takes only a synthetic, generalised view of the evidence from several sites at once. It is therefore concerned only with trends in the faunal data, with similarities and differences across time and space. It is not concerned with reconstructing putative livestock populations and husbandry strategies at individual sites. This comparative, generalising approach is crucial. It aims to discern patterns in animal husbandry which characterise the regions and periods under study, and which transcend the quirks, outliers and biases in individual assemblages. With this in mind, the comparative analyses presented below address two key issues: (i) the importance of sheep relative to other livestock species; and (ii) the importance of wool relative to other animal products, between the fifth and ninth centuries.

The importance of sheep

In discussing the first issue, this chapter follows in a long and respectable tradition of comparing the percentages of cattle, sheep and pig bones in zooarchaeological assemblages. Such analyses have grown much more comprehensive as the national faunal dataset has expanded. In 1976, Clutton-Brock had data for only five Anglo-Saxon assemblages; in 2014, Holmes assembled data from 315 (Clutton-Brock 1976; Holmes 2014). Archaeological and historical evidence support the universal notion that these three species have been the mainstay of British animal husbandry throughout the Iron Age, Roman and medieval periods: the cow (*Bos taurus* L.), the sheep (*Ovis aries* L.), and the pig (*Sus domesticus* Erxleben). While other livestock species were undoubtedly present in Anglo-Saxon England – poultry, for instance, and horses – they are usually much less in evidence. In the case of bird bones, this archaeological dearth is partly due to their fragility and, in consequence, the lower probability of their survival, recovery and identification. In any case, cattle, sheep and pig bones all share two attributes – archaeological ubiquity and relative robustness – which justify their juxtaposition in these three-way faunal comparisons.

The only real bone of contention here is the use of 'sheep' as a category, since sheep and goat skeletal remains are scarcely distinguishable but for their horncores. It is therefore typical for zooarchaeological studies to employ an umbrella class, variously glossed as 'sheep/goat', 'ovicaprid', 'caprine', or similar terms, with only a few bones specifically assigned to either sheep or goat (*Capra hircus* L.). This approach is sound and scientifically responsible. In practice, however, there are exceptionally few positively identified goat remains from Anglo-Saxon England. Where specific identifications are possible, they most often specify sheep. Goats appear genuinely to have been relatively rare in Anglo-Saxon farming (Banham & Faith 2014, 95–97). In this book, therefore, I make a working assumption that ambiguous 'ovicaprid' bones are most likely to represent sheep unless otherwise stated.

With this in mind, the basic method has been to calculate percentages of cattle, sheep (ovicaprid) and pig bones for each phase at each site, in as much detail as could be obtained from available reports. The vast majority of these bones, as is common for large mammals, were recovered by hand (rather than through soil-sieving) which will have lent a slight bias towards cattle, as the species with the largest, most robust bones (O'Connor 2011, 364). Beyond calculating percentages, I have not transformed the bone-counts reported by zooarchaeologists. All reported counting methods were variations of the NISP (Number of Identified Specimens) method, where the eponymous 'specimens' are usually individual bones. While the NISP approach does admit subtle variations, such as in the treatment of articulated skeletal elements or in the selection of which bones to count, the essential criterion is that cattle, sheep and pig bones have been treated the same way *within* each assemblage, so that it is legitimate to calculate their relative proportions on a site-by-site basis. Once again, cattle bones gain a slight advantage through the NISP approach, by virtue of their greater size and osteological resilience. Alternative quantification methods, such as MNI (Minimum Number of Individuals), aim to address this bias and other deficiencies in NISP. However, these alternative counts do not lend themselves well to statistical manipulation or inter-site comparisons (O'Connor 2000, 54–67). It is sufficient here to expect, from the outset, that there is a likely bias towards higher proportions of cattle bones in this dataset.

One final methodological point deserves consideration: small and large assemblages are not necessarily directly comparable. Larger assemblages are inherently more likely than smaller assemblages to be representative of their original (deadstock) populations, but there is no obvious, objective threshold to distinguish large from small (Amorosi *et al.* 1996). The fact that cattle bones account for 56% of the Anglo-Saxon assemblage from Wickhams Field (Berkshire), for instance, is hardly significant when one considers that there are only nine bones from this phase (Hamilton-Dyer in Crockett 1996, 157–159). A popular cut-off point is a quorum of 100 bones, which at least renders the species percentages more meaningful. For instance, Sykes excludes assemblages with fewer than 100 cattle, sheep and pig bones in total (Sykes 2007, 9). Hambleton, however, presents an intriguing argument for a quorum of 300 bones in her study of Iron Age animal husbandry. Above this level, she points out, there are fewer discrepancies between NISP and MNI ratios. This observation could be thought to indicate that these assemblages are more representative, large enough to transcend some of the methodological biases inherent in NISP and MNI (Hambleton 1999, 39). Whatever quorum one chooses will necessarily be arbitrary. A minimum total of 300 bones is preferred here simply as a more rigorous criterion that still produces a dataset large enough to analyse: to wit, 62 assemblages spanning the fifth to ninth centuries, widely though not evenly scattered across the case study regions (Figure 35).

Such are the necessary caveats; now we may examine the dataset. The relative proportions of cattle, sheep and pig bones in the various assemblages are plotted below on a triangular graph, the traditional means of illustrating such ratios (Figure 36). These graphs can be difficult to read, but the essential pattern here should be clear: the relative proportions of cattle, sheep and pig bones remain fairly similar across most assemblages. In accordance with the predicted bias towards bovine skeletons, cattle bones tend to be dominant, constituting around 25–65% of the total, whereas sheep bones usually make up 20–50%, and pig bones typically 5–35%. These percentage ranges describe 41 of the 62 assemblages, and several of the remaining assemblages differ only slightly from this overall trend. However, there are also some distinct outliers. I have argued above that this study should concentrate on the general rather than the exceptional, so as not to be misled by assemblages' individual biases. Nevertheless, it is worth examining the outliers to investigate whether they might conform to a pattern of their own: whether there is a common factor, such that the exceptions may in fact prove a rule.

Two outlying assemblages have outstandingly high proportions of pig bones: those representing successive seventh and eighth- to ninth-century phases at Wicken Bonhunt (Essex). One assemblage has a uniquely high proportion of sheep bones: that representing post-Roman activity at Uley (Gloucestershire), a ritual site used throughout the Iron Age, Roman and post-Roman periods. Finally, four assemblages have especially high proportions of cattle bones: they derive from fifth- to sixth-century contexts at Heybridge (Essex), Spong Hill (Norfolk), Latimer (Buckinghamshire), and the Beech House Hotel site in Dorchester-on-Thames (Oxfordshire). Certainly, the environment and topography of these sites are ecologically compatible with their osteologically dominant fauna. Wicken Bonhunt's clayey hinterland is thought to have been

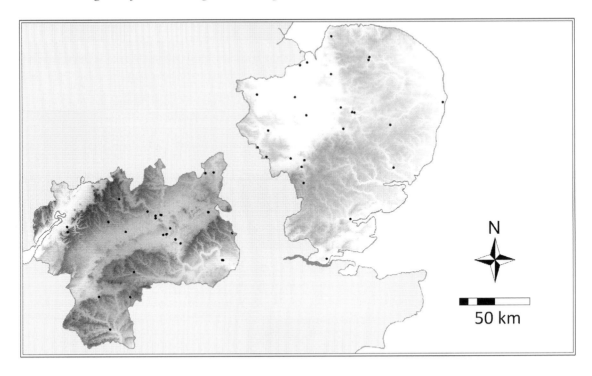

FIGURE 35. Distribution of sites with faunal assemblages with at least 300 cattle, sheep and pig bones, within the study regions

well-wooded in the Anglo-Saxon period and therefore rich in pannage for pigs (see Chapter 1). For Uley, the chalk uplands of the Cotswolds are well-suited to grazing sheep. Heybridge, Spong Hill, Latimer and Dorchester-on-Thames are all close to lowland valley pastures apt for grazing cattle. But such correlations alone fail to provide a satisfactory explanation, simply because almost all of these locations are shared by other assemblages which are *not* especially dominated by the bones of a particular species (the exception is Wicken Bonhunt, which provides the only Anglo-Saxon faunal assemblage from the clays of north Essex).

What all of these outlying assemblages (except for those from Wicken Bonhunt) do have in common, however, is a clear Romano-British precedent. The bones of sheep and/or goats dominate all preceding phases at Uley, and cattle bones dominate late Roman assemblages at Heybridge, Spong Hill, Latimer and Dorchester-on-Thames. There is a case, therefore, for seeing in these assemblages the last flowering of a late antique phenomenon: the twilight products of Roman Britain's rural economy, lingering at least into the fifth century, and possibly the sixth. Perhaps they represent an increasingly redundant system of centralised culling and butchery, a relic of the Roman market economy. Indeed, the exceptional assemblages from Wicken Bonhunt – exceptionally late among the outliers, being seventh-century and later in date, and exceptionally porcine in composition – might reflect comparable practices in the revived exchange economy of the long eighth century. The pig bones themselves are unusually dominated by skulls and jaws at Wicken Bonhunt, which Crabtree persuasively takes to mean that the main meat-bearing bones

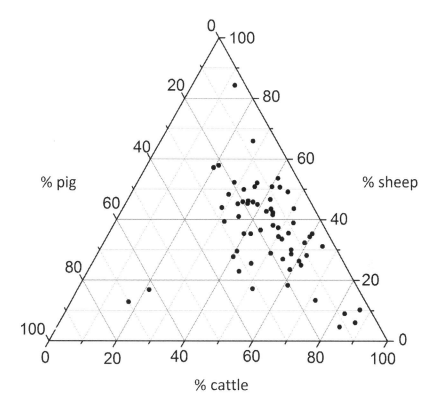

FIGURE 36. Tripolar graph showing the relative proportions of cattle, sheep and pig bones in assemblages in the project dataset

had been exported elsewhere, i.e. that Wicken Bonhunt was a specialist pork-producing settlement (Crabtree 2012, 16–17, 26–27).

These few outlier assemblages might thus prompt a hypothesis. The apparent 'specialisation' in cattle or sheep in a few post-Roman assemblages could reflect specialised, market-based *mobilisation* rather than specialised *husbandry* of a particular species. Conversely, the general homogeneity of cattle-sheep-pig ratios in most Anglo-Saxon faunal assemblages need not reflect a complete lack of specialised animal husbandry, but rather the lack of specialised meat markets linking a network of villas (such as Latimer), small towns (such as Heybridge and Dorchester-on-Thames) and ritual centres (such as Uley). Such a model would be consistent with Crabtree's observation that butchery marks appear more standardised on Romano-British animal bones than on their post-Roman counterparts (Crabtree 2012, 8–12).

So even if a particular region or period was marked by an emphasis on sheep husbandry in Anglo-Saxon farming, we need not expect any *individual* assemblage from that region and period to be dominated by sheep bones. Rather, we might look for a slighter but more sustained concentration of sheep bones across *several* assemblages, in the absence of a meat market system to create heavily-dominated nuclei at a few specific sites. Within the project dataset, however, the chronological patterns are very slight indeed: an extremely subtle fall in the proportions of cattle bones between the fifth and ninth centuries, and a correspondingly subtle rise in the proportions of both sheep and pig bones over the same period (Figures 37 to 39). These very modest trends are arguably well within the margin of error for this kind of analysis, and it would be a highly optimistic analyst who inferred an eighth-century wool boom from these data.

A geographical pattern, on the other hand, can be discerned more clearly. Again excluding the outliers so as to illustrate trends in the underlying mass of data, Figures 40 to 42 present the same faunal data as interpolated maps. Much as a cartographer interpolates from a series of elevation readings to produce a contoured terrain map, so have I interpolated the relative abundance of cattle, sheep and pig bones across the study regions from the series of assemblages in the project dataset. The specific method applied is Inverse Distance Weighting, which takes the distances between data points into account when interpolating overall patterns (Chapman 2006, 76). Ideally, the individual 'readings' (i.e. assemblages) would be evenly distributed across the regions to ensure that each area is equally well represented. Since this is not the case, these interpolations should be read in tandem with Figure 35 above, which shows the complete distribution of all assemblages in the dataset. One advantage of this approach to mapping NISP data is that it can still display data from closely adjacent assemblages, or even overlapping assemblages representing different phases at the same sites. Such overlapping data-points would obscure each other in, for example, a map of pie charts representing NISP ratios (as attempted in my thesis: McKerracher 2014a, 150).

What, then, do these interpolated maps reveal? While there are some apparent 'hotspots' with relatively high proportions of cattle or pig bones, these tend to represent individual sites rather than reflecting more general geographical patterns (Figure 40 and Figure 42). For sheep bones, however, there seems to be a genuine regional concentration encompassing the basin of the Great Ouse, the Wash, the Fens and the Breckland (Figure 41). While one could of course wish for more data-points, this pattern is at least sustained across several assemblages in a clearly defined geographical region, centred on the Fens. The ecologically (or teleologically) minded historian will not be surprised by this pattern. The lighter soils of the Breckland and East Anglian Heights are well-suited to sheep grazing (Williamson 2003, 79), and the wealth of later medieval wool merchants was to be founded partly in 'lush fen summer pastures', enriching fen-edge market towns like Peterborough (Pryor 2010, 370).

In sum, there is only very slight zooarchaeological evidence for a shift in emphasis towards sheep husbandry in the seventh to ninth centuries in the study regions, although there may well have been a regional tendency towards sheep husbandry in and around the Great Ouse basin, and an inverse tendency towards cattle elsewhere in the study regions. A specialised focus on pig husbandry, and perhaps even specialist pork production, may have been a highly localized innovation of the seventh to ninth centuries on the relatively well-wooded clay plateaux of north-west Essex. While we should not read too much into the evidence of a single site here (Wicken Bonhunt), it is nonetheless

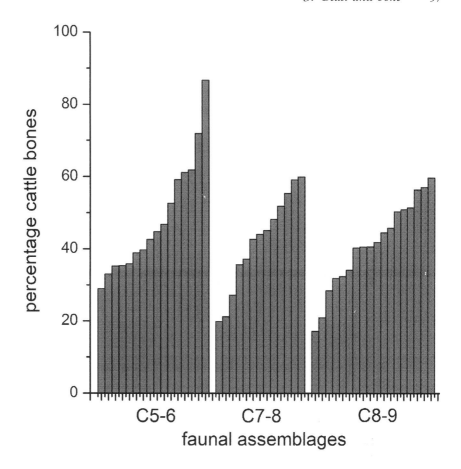

FIGURE 37. Percentages of cattle bones in faunal assemblages, by period, excluding outliers as discussed in the text

interesting to note that a closely-contemporary assemblage from St Albans Abbey (Hertfordshire), which occupies a somewhat similar geological position between chalk uplands and London Clay, is likewise relatively rich in pig bones (Figure 43; Crabtree 2012, 14–15; McKerracher 2016a).

The importance of wool

So much for the sheep; what can be said of the wool? Zooarchaeologists have long attempted to deduce what animal products were being produced at a site from its surviving faunal remains. They have sometimes developed detailed, idealised models of livestock populations and kill-patterns representing, for example, meat-focused or milk-oriented herds of sheep or cattle (Payne 1973 being a seminal example). These models are based on the premise that farmers will control the demographic profiles of their livestock – the ratios of old to young, male to female, castrated to entire – to ensure the efficient return of the desired animal products, or else well-balanced subsistence. In practice, however, these models are difficult to apply directly to zooarchaeological data. This is

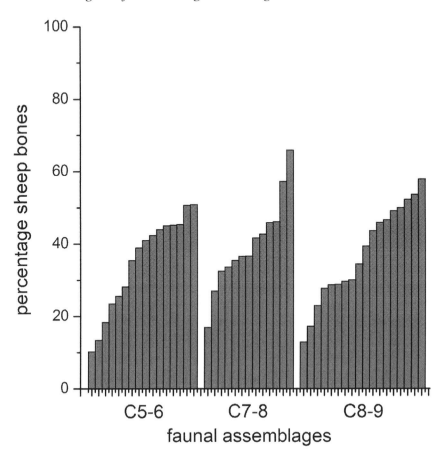

FIGURE 38. Percentages of sheep bones in faunal assemblages, by period, excluding outliers as discussed in the text

partly because of the wide taphonomic gulf separating ancient populations from modern specimens, but also because we cannot necessarily assume that ancient farmers were always economically rational in a modern sense. In the freedom of his blog, O'Connor has warned zooarchaeological novices that 'most animal husbandry decisions are the result of collective beliefs that have no basis in fact combined with individual cantankerousness. Optimal productivity is an occasional, and largely inadvertent, by-product' (https://osteoconnor. wordpress.com/2015/07/13/useful-advice-for-newbie-zooarchaeologists, accessed April 2017).

Such caveats do not, however, completely undermine the study of sexing and mortality profiles of faunal assemblages. Instead, they should encourage us to look for simpler but stronger patterns, and to focus on excluding the unlikely rather than proving the possible. For instance, if, in a given assemblage, a clear majority of ageable cattle bones represent beasts that were elderly when slaughtered, it seems most unlikely that prime beef was the main animal product sought at that site, since maintaining cattle for a greater number of years past bodily maturity does not increase the volume or quality of meat

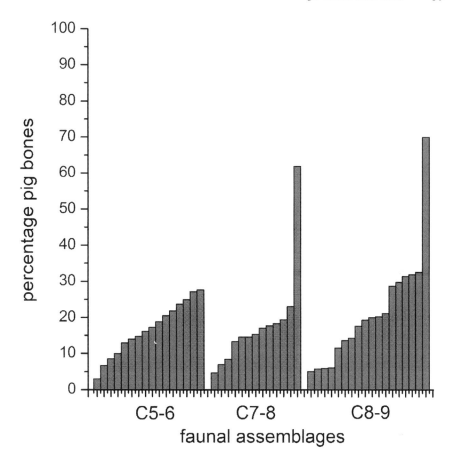

FIGURE 39. Percentages of pig bones in faunal assemblages, by period, excluding outliers as discussed in the text

that can be gained when they are slaughtered. Conversely, a site at which most sheep were culled in their first year of life is unlikely to have been targeting wool production, since a single fleece would be a poor return for the effort of raising a lamb to maturity.

With these considerations in mind, I have collected and studied mortality (age-at-death) data and male-to-female ratios, where available, for the faunal assemblages in the project dataset. To enhance the comparability of the ageing data, I have focused exclusively on mandibular wear, an approach which is based on the consistent, age-related sequence in which permanent teeth erupt in the lower jaw and then wear down from enamel onto dentine. While other ageing methods are available, mandibular wear is particularly useful, first because mandibles and teeth are relatively robust and often survive better than other skeletal elements, and secondly because teeth grow and wear in a known sequence that spans an animal's entire life, not just its youth. The dental sequence is variable enough (for genetic or dietary reasons, for example) that we should hesitate to apply precise ages to mandibles; but it is nonetheless consistent enough that we can reliably differentiate the old from

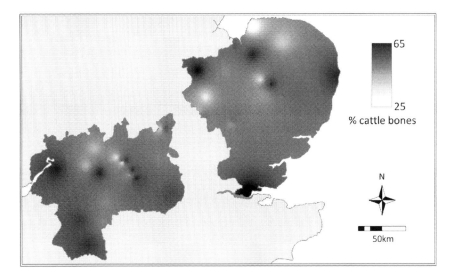

FIGURE 40. Interpolated map of the proportions of cattle bones in assemblages within the study regions, excluding outliers as discussed in the text

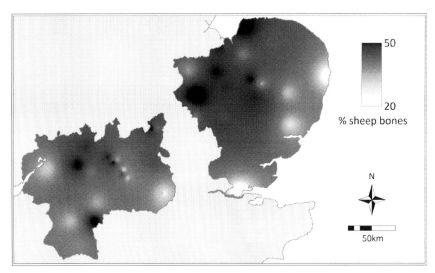

FIGURE 41. Interpolated map of the proportions of sheep bones in assemblages within the study regions, excluding outliers as discussed in the text

the young, and the young from the juvenile. Mandibular wear data for the assemblages in the project dataset are standardised according to O'Connor's five age 'milestone' categories of Juvenile, Immature, Subadult, Adult and Elderly. Broadly speaking, individuals slaughtered while Immature or Subadult are more likely to have been kept for primary products, while those kept alive until Adult or Elderly are more likely to have been kept for secondary products.

I have required each assemblage to have at least 30 aged mandibles per species, a low and arbitrary quorum, but one which at least excludes very small (and therefore potentially very unrepresentative) mandibular sequences. On this basis, there are seven sequences for cattle and 17 for sheep. Sexing data were much rarer and more variable in their derivation methodologies, and

FIGURE 42. Interpolated map of the proportions of pig bones in assemblages within the study regions, excluding outliers as discussed in the text

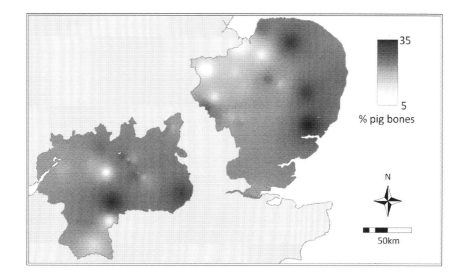

FIGURE 43. Locations of Wicken Bonhunt and St Albans Abbey

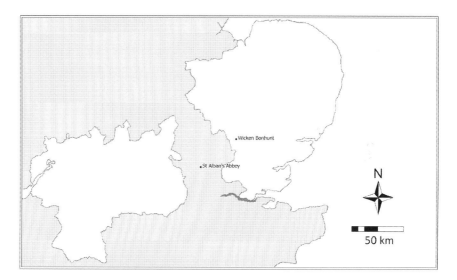

therefore practically impossible to standardise. Such meagre datasets do not lend themselves well to detailed statistical analysis or interpolated mapping, but are small enough simply to examine by eye. For easy visual assessment, therefore, I have presented the mandibular sequences for sheep and cattle in Table 1 and Table 2 respectively, giving the percentage of animal bones belonging to each of O'Connor's milestone categories. Sites yielding mortality or sexing data are mapped in Figure 44. I do not consider data for pigs in this discussion, since there is no real debate over which products pigs were raised for at any site, in any period: pork, fat, and piglets (McKerracher 2016a).

So what patterns exist within these data? First, there is a distinct group of assemblages whose sheep mandibles are dominated by Adult and/or Elderly specimens. The clearest examples are from Bloodmoor Hill, Brandon, Ely and

Site	Total	% Juvenile	% Immature	% Subadult	% Adult	% Elderly
Barrow Hills	71	11.3	15.5	19.7	53.5	0.0
Barton Court Farm	65	6.2	23.1	30.8	38.5	1.5
Bloodmoor Hill	38	0.0	0.0	7.9	86.8	5.3
Brandon	410	3.7	7.6	12.7	73.9	2.2
Collingbourne Ducis	43	16.3	34.9	16.3	18.6	14.0
Consortium Site, Ely	56	5.4	5.4	7.1	76.8	5.4
Dorney	85	1.2	18.8	18.8	55.3	5.9
Godmanchester	70	7.1	5.7	27.1	15.7	44.3
Harston Mill	100	6.0	23.0	21.0	23.0	27.0
Ipswich	117	5.1	35.9	10.3	48.7	0.0
Market Lavington	40	17.5	20.0	17.5	45.0	0.0
Pennyland	92	4.0	32.6	24.6	38.8	0.0
Redcastle Furze	44	6.8	27.3	27.3	38.6	0.0
Walpole St Andrew	40	0.0	20.0	30.0	50.0	0.0
West Stow	1293	6.1	38.5	18.0	36.1	1.2
Wicken Bonhunt	108	9.3	8.3	4.6	76.9	0.9
Wolverton Turn	45	6.7	8.9	33.3	48.9	2.2

TABLE 1. Sheep mortality profiles from mandibular sequences in the project dataset

Wicken Bonhunt (Table 1). Similar but much slighter trends could be identified, perhaps, at Godmanchester and Dorney, but less persuasively. At Ramsbury too, although quantified data were unavailable for analysis, the 50 sheep mandibles are said to be dominated by mature individuals; and Noddle, using an idiosyncratic ageing method that is difficult to standardise, observed the same tendency towards ovine maturity at North Elmham Park (Coy in Haslam *et al.* 1980, 47; Noddle in Wade-Martins 1980, 396). All of these assemblages are certainly of seventh- to ninth-century date except for that from Bloodmoor Hill, whose occupation sequence is thought to span the sixth to eighth centuries. A seventh century or later date for many of the Bloodmoor Hill mandibles is therefore possible, but not certain.

Of the remaining sheep mortality profiles, most are theoretically consistent with an unstructured, unspecialised cull, as in the model suggested by Crabtree for West Stow (Crabtree 1989). Variations upon this basic model would be equally applicable to such chronologically and geographically diverse assemblages as those from Redcastle Furze, Pennyland, Harston Mill and Barton Court Farm, none of which demonstrates clear signs of specialised management or structured culling. There is, however, some indication of an early cull (a small peak of Immature individuals) at Collingbourne Ducis, whose assemblage is predominantly of seventh- to ninth-century date. The mandibular sequence here was taken by the original analyst to imply 'an interest in meat rather than wool' (Hamilton-Dyer in Pine 2001, 106). At Wolverton Turn, whose assemblage is again predominantly of seventh- to ninth-century date, there is a joint emphasis upon Subadult and Adult individuals, which again could indicate some economic emphasis upon meat production beyond that evidenced at most

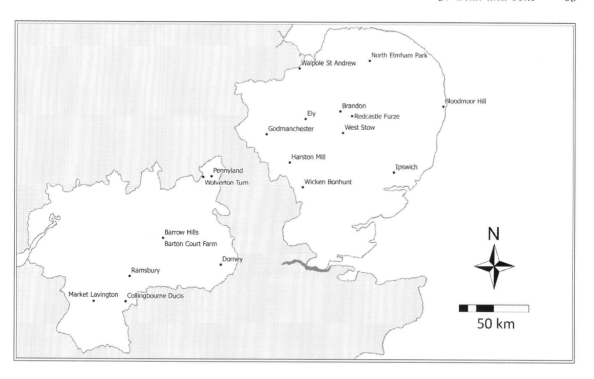

FIGURE 44. Map of sites with mortality and/or sexing data in the project dataset

other sites of the period (Sykes in Preston 2007, 108). A similar interpretation might be offered for Pennyland's sixth- to eighth-century assemblage (Holmes in Williams 1993, 150).

Sexing data shed a little further light on what sheep products may have been pursued at some of these settlements. Brandon's sheep are dominated by wethers (castrated males), which are thought to produce the best wool (Crabtree 2007, 164–167). By contrast, females appear to dominate at Wicken Bonhunt, in a pattern perhaps more consistent with dairying or breeding stock; the Godmanchester and Dorney assemblages are similarly dominated by females. At North Elmham Park, by contrast, ewes and wethers exist in roughly equal proportions, a pattern which could reflect the importance of both milk and wool at that site (Crabtree 2012, 49; Baxter in Gibson 2003, 192; Powell & Clark in Foreman *et al.* 2002, CD-ROM; Noddle in Wade-Martins 1980, 396). An apparent dominance of rams is seen only, and appropriately, at Ramsbury (Coy in Haslam *et al.* 1980, 47).

Amongst the cattle bones, a noticeably 'young' cull is evidenced only in Pennyland's sixth- to eighth-century assemblage, while more mature mandibles dominate the seventh- to ninth-century assemblages from Brandon and Wicken Bonhunt and the sixth- to eighth-century sequence from Bloodmoor Hill (Table 2). Female cattle are found to dominate at Brandon and Dorney, males at Wicken Bonhunt (Crabtree 2012, 31–46; Powell and Clark in Foreman *et al.* 2002, CD-ROM).

Site	Total	% Juvenile	% Immature	% Subadult	% Adult	% Elderly
Bloodmoor Hill	69	0.0	15.9	17.4	44.9	21.7
Brandon	83	1.2	2.4	15.7	63.9	16.9
Dorney	128	7.8	12.5	28.9	41.4	9.4
Ipswich	34	0.0	14.7	26.5	50.0	8.8
Pennyland	34	17.6	17.6	44.1	20.6	0.0
West Stow	290	13.8	17.9	23.4	9.7	35.2
Wicken Bonhunt	99	1.0	3.0	9.1	58.6	28.3

TABLE 2. Cattle mortality profiles from mandibular sequences in the project dataset

How should we interpret these trends? First, we may observe that there is no evidence for any kind of structured culling pattern – or imbalanced sex ratios – necessarily earlier than the seventh century. Hence, there are no strong empirical grounds for rejecting the popular theory, articulated most fully by Crabtree, that animal husbandry in the fifth and sixth centuries was characterised by a broad-based strategy of self-sufficiency, with no marked specialisation in any primary or secondary product (Crabtree 1996; 2010, 131; 2012, 57–59). Second, we may observe that even from the seventh century onwards, there is no strong evidence for targeted beef, mutton or lamb production (i.e. no relatively young culls) in East Anglia or Essex, whereas there is a little, scattered, possible evidence further west in the study regions at Pennyland, Wolverton Turn and Collingbourne Ducis. Third, there is an inverse pattern whereby the only strong evidence for an emphasis upon *secondary* products is *within* East Anglia and Essex (Dorney in the Middle Thames valley being the sole, debateable exception). However, there are no clear regional patterns in the sexing evidence, which is unsurprising given the overall dearth of such data. The apparent dominance of castrates at Brandon, and the possibility that this raises of a specialised wool flock, makes an intriguing individual case study and is consistent with the mortality data indicating a focus upon secondary products; but there is no wider trend concerning wethers discernible within the project dataset.

Finally, there is a complementary dataset of cattle and sheep pathologies: specifically, arthritic changes in limb bones which are sometimes thought to have agricultural significance. Detailed analysis of these pathologies would require a specialist study. Here, I have simply recorded the occurrence of such evidence, where noted in the original zooarchaeological reports. The evidence falls into two categories: 'penning elbow' and 'traction pathologies'. Traction pathologies are arthritic changes to cattle long bones which, while sometimes occurring simply because of old age, can be symptomatic of heavy labour, i.e. the use of cattle as beasts of burden to pull ploughs or carts (Bartosiewicz *et al.* 1993; 1997). The recorded instances in the dataset are variously described as exostosis, eburnation, splaying and lipping of metapodia and phalanges; fusion of articulating bones; and generic wear on joints. The term 'penning elbow' denotes lipping or exostosis of sheep humeri and radii. Again, it is an age-related condition, but according to

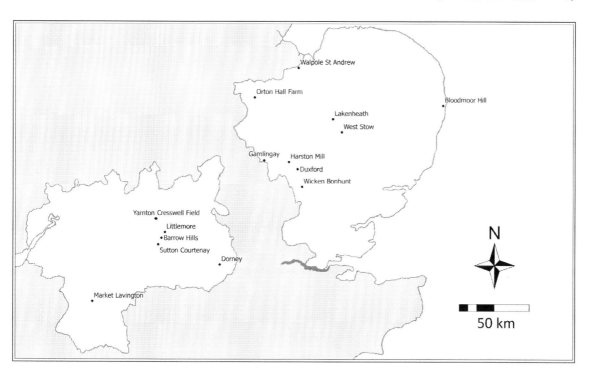

FIGURE 45. Sites at which cattle traction pathologies may be evident

some scholars can be exacerbated by 'rough handling or confinement or penning' (Poole in Davies 2008, 178; Dobney *et al.* 2007, 184–186).

Occurrences of these two pathologies in the project dataset are mapped in Figure 45 and Figure 46. It appears that there are no clear regional patterns among these data. The only possible chronological pattern is that no instances of 'penning elbow' need be any earlier than the seventh century, which would be consistent with an increase in the maturity and confinement of flocks at this time. Overall, however, the mere occurrence of these pathologies adds little to the overall emerging picture of animal husbandry in this period.

Conclusions

Despite the cautious, rather sceptical approach taken in this chapter, some significant conclusions may be drawn. Cattle, sheep and pigs were ubiquitous throughout the fifth to ninth centuries in the study regions. Their relative economic importance is unclear, but does not appear to have changed greatly during these centuries; little can be read into an extremely slight shift away from cattle and towards sheep and pig bones. Animal husbandry around the Fens, Breckland and Great Ouse basin may have been particularly sheep-oriented through the fifth- to ninth-century period. More widely in East Anglia and Essex, there are signs that both cattle and sheep were kept to greater ages more often from the seventh century onwards, suggesting a greater abundance of

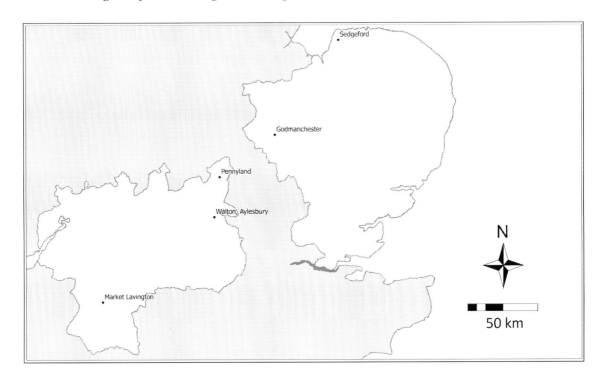

FIGURE 46. Sites at which sheep 'penning elbow' pathologies may be evident

secondary products. An inverse trend towards more primary products further west is very slight. In fact, the strongest evidence for specialised meat production is in the unique pig bone assemblage from Wicken Bonhunt in Essex, dating from the seventh century onwards: the first suggestion of anything like a market for meat since the immediate post-Roman period.

Besides the question of secondary products, there are two further implications arising from the growing maturity of East Anglian livestock mortality profiles in the long eighth century. First, farmers were clearly capable of overwintering much of their livestock by this period, or else such high proportions of Adult/ Elderly beasts would not be evident. Second, since smaller proportions of individuals were being culled at younger ages, the overall numbers of cattle and sheep are likely to have increased: higher life expectancy implies overall demographic growth unless birth-rates are correspondingly checked (by reducing breeding stock or killing neonates). It would be very difficult to identify a reduction in breeding in the archaeological record, whereas there is independent evidence to support the argument for both large livestock populations and effective overwintering – namely, the evidence for paddocks, droveways and hay meadows discussed in Chapter 2. Indeed, the hypotheses proposed in Chapter 2 seem highly compatible with the ideas raised in this chapter. There, I suggested that the encroachment of spreading arable would have restricted the availability of outlying pasture from the seventh century onwards; now we might add that growing livestock populations could also have increased pressure on diminishing

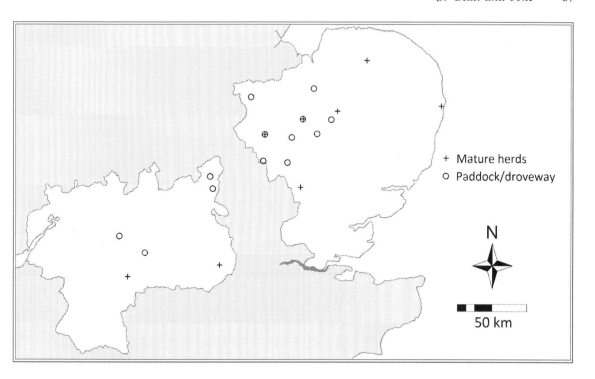

FIGURE 47. Distributions of mature herds and droveway/paddock complexes in the study regions

grazing land. There, I suggested that hay meadows began to be managed at this time to compensate for the loss of winter pasture or wood-pasture, and so to feed overwintered livestock closer to the farm; now we can see from the faunal evidence that more cattle and sheep were being successfully overwintered at this time, and fewer were being slaughtered at younger ages. In short, higher proportions of livestock were seeing more winters.

The geographical distributions of these different phenomena show some overlap, but not in a straightforward fashion. For instance, the area around the Fens, fen-edge and wider hinterland sees a coincidence of higher proportions of sheep bones, increasing maturity of livestock, and a concentration of droveway and paddock systems (Figure 41 and Figure 47). Yet the increasing maturity of livestock is a trend that embraces East Anglia and Essex more widely: Bloodmoor Hill and Wicken Bonhunt are far removed from the fenland, and occupy different environments both from the fen-edge and from each other. The droveway and paddock complexes, meanwhile, although they are concentrated around the fen-edge, extend far to the west across the Midland Clays and into the Upper Thames valley, where there is no clear evidence of increasing herd maturity. While adult sheep predominate in the assemblage from Yarnton, for instance, there are fewer than 30 aged mandibles in that assemblage, and the data are presented for the whole Anglo-Saxon period, such that we cannot closely date the development between the fifth and eleventh centuries (Booth *et al.* 2007, 336).

Perhaps this overall pattern thus reflects the coincidence of three regional trends: the East Anglian trend towards greater herd maturity meeting the Upper Thames/Great Ouse propensity for paddocks and droveways, in the sheep-oriented environs of the Fens and fen-edge. On the other hand, it is not unlikely that the geographical discrepancies are artefacts of taphonomy. There are too few mandibular sequences in the dataset for absences to be very meaningful, while the recognition of paddocks and droveways has involved nice, arbitrary judgements which need not be exclusive or definitive. Others might consider the extensive parallel ditches at North Elmham Park and Wicken Bonhunt to be part of the same phenomenon, thus extending its known reach deeper into East Anglia and Essex. In this case, the true significance of the convergence of data around the fenland basin could be in demonstrating the close relationship between these disparate strands of evidence which survive only singly elsewhere: mature herds, paddocks and droveways, related to closer spatial management of livestock in response to a reduction in grazing land.

Throughout this chapter and the last, I have repeatedly adduced a growth in arable as an explanatory factor behind settlement expansion and loss of pasturage. So far, this arable growth has been more or less mythical. Only the very slim (and generally late) evidence for heavy ploughing has borne any direct relation to crop husbandry. It is time to address this imbalance, and consider in detail the evidence for the growth of arable farming.

CHAPTER 4

The growth of arable

In the public imagination, Anglo-Saxon England engenders thoughts of skirmishes before cereals, swords before ploughshares: the complete warrior society. Anglo-Saxon histories for a general readership typically feature cover images such as Alfred in warrior garb, the Sutton Hoo helmet, or items from the Staffordshire Hoard – popularly perceived as war booty and certainly containing élite weaponry. It remains to be seen whether the Lyminge plough coulter, equally rare and significant in its own way, will achieve such celebrity. The literary legacy of the Anglo-Saxons has of course contributed to this bias, favouring fighting over farming or even eating. When the heroes of *Beowulf* are feasting, we never learn what food is on the table (Banham 2004, 8).

On the other hand, the rhythms of the *Anglo-Saxon Chronicle* and other early medieval annals are only partly defined by a succession of battles. Harvests, famines and cattle mortality are also charted (Banham & Faith 2014, 108; Wood 2010, 61–62). The warrior society is certainly an important concept in understanding Anglo-Saxon culture, but early medieval England was primarily a society of farmers whose fortunes were tied more fundamentally to the cornfield than to the battlefield. The Old English lord was the bread keeper, OE *hlaford*, his lady the dough-kneader, OE *hlafdige* (Banham 2004, 17). Whatever the changing patterns of animal husbandry (discussed in the previous chapter), cereals are likely to have made the supreme calorific contribution to Anglo-Saxon diets (Banham 2004, 13). It is natural to ask, therefore, what part crop husbandry played in the transformation of farming between the seventh and ninth centuries. This book has focused so far upon pastoral farming. Even the discussion of farms and fields concluded that paddocks, droveways and perhaps hay meadows were a more marked innovation of the period than were new arable field systems (see Chapter 2). Yet the notion of greater control over livestock also implies that there was land – arable land – which needed protection from the hungry beasts thus enclosed. Hence it is possible that the innovations identified in animal husbandry were not solely concerned with delivering animal products, but also with allowing the growth of arable farming. What evidence exists, then, for the increase of arable surpluses in the long eighth century? This chapter considers settlement evidence for crop storage and processing (mills, ovens, barns and granaries), traces of arable farming in the wider environmental record, and finally the testimony of the remains of the crops themselves, as recovered from archaeological excavations.

Settlements and structures

It is widely noted that late Roman Britain produced substantial crop surpluses, not only to sustain military garrisons and urban populations, but also for export to the Rhineland (Mattingly 2006, 501). The storage and processing of these surpluses made a distinctive impact on Romano-British rural settlements, as large new structures were erected to fulfil these functions: barns, granaries, watermills and grain ovens, the latter used to dry harvested crops and to malt grain for brewing (see, for example, Fleming 2010, 17; van der Veen & O'Connor 1998). I suggested in the last chapter that late Roman systems for mobilising livestock, perhaps in a meat-market network, in some places enjoyed an extended twilight into the fifth and possibly even sixth centuries (see Chapter 3). Something similar appears to have happened with these structural elements of the arable economy. This is not especially surprising. Mounting evidence from pollen and landscape studies shows that the post-Roman period in much of Britain witnessed neither woodland regeneration nor the loss of field boundaries. Thus the framework of British farming, if not its finer details, appears to have survived the withdrawal of Roman governance (Dark 2000, 150–154; Upex 2002; Rippon *et al.* 2015). What the structural evidence adds to this impression of landscape survival is a hint that large surpluses – large enough to require specially-built facilities – were still being mobilised after imperial rule had ended (Figure 48). Hence an excavated Roman watermill complex at Ickham, Kent, has produced artefactual evidence suggesting continuity of operation into the fifth century (Bennett *et al.* 2010, 70–71, 326). Isolated Romano-British millstones are more commonly found than are entire watermills, and some of these too have been identified in apparently post-Roman contexts, their size and shape indicating mechanised mills rather than hand-quern usage (Wilson in Bennett *et al.* 2010, 63–64). In the study regions, for example, there are supposed millstone fragments in the south-east paddock at Barton Court Farm (Oxfordshire), believed to be of fifth-century date, and in a fifth- to sixth-century context at Brettenham, Norfolk (Spain in Miles 1986, microfiche 5:A13; Mudd 2002, 76–77).

Evidence for post-Roman barns or granaries is more equivocal. There is nothing distinctively granary-like about a post-Roman structure at Latimer, Buckinghamshire, originally suggested to be a granary by the excavator (*pace* Branigan 1971, 187). However, a more uniform nine-post structure at Orton Hall Farm, Cambridgeshire, in that site's difficult post-Roman phase (fifth- to sixth-century), has more convincingly been seen as a crop-storage structure (Mackreth 1996, 89–91). The key difference between these two structures is in the configuration of their posts – uniform at Orton Hall Farm, irregular at Latimer – which has specific relevance for storage technologies (see further below). There is also a so-called 'barn' at Orton Hall Farm which is said to have survived into the post-Roman phase, including a large stone-built grain oven within its walls (Mackreth 1996, 75–78). Other similar ovens elsewhere in the study regions have likewise produced evidence for fifth-, if not

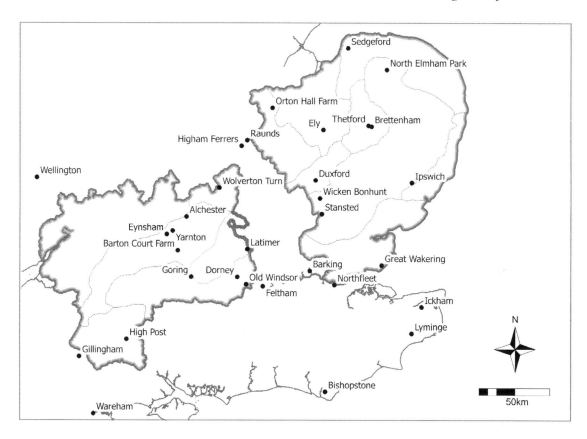

Sedgeford

North Elmham Park

Orton Hall Farm

Ely Thetford Brettenham

Higham Ferrers Raunds

Wellington

Duxford

Wolverton Turn Wicken Bonhunt Ipswich

Alchester Stansted

Eynsham
Yarnton
Barton Court Farm Latimer

Great Wakering

Goring Dorney Barking
Old Windsor Northfleet
Feltham

Ickham

High Post Lyminge

Gillingham

Bishopstone

Wareham

N

50km

FIGURE 48. Map of sites
discussed in this section

sixth-century survival. Three of these exemplify the widespread and well-known 'T' shape. At Alchester and Goring (Oxfordshire) and High Post (Salisbury, Wiltshire) are Roman ovens whose post-Roman afterlife is implied by their stratigraphic association with fifth- to sixth-century pottery (Booth *et al.* 2001, 202–207; Allen *et al.* 1995, 39–44; Powell 2011, 29–33). A more unusual example, the Roman 'drying building' identified at Duxford (Cambridgeshire) with a curving C-shaped flue, offers no direct artefactual evidence of post-Roman use, but Anglo-Saxon features from a later phase of occupation appear to respect its position and alignment, suggesting at least its continued visibility (Lyons 2011, 91). There is no easy way of assessing, however, how far such exceptional archaeological finds represent a wider historical phenomenon. For the vast majority of Romano-British grain ovens, granaries, barns and watermills there is no evidence of usage after *c.* AD 410, but such evidence would always be difficult and ephemeral: crop storage and processing are not readily dateable activities.

What we *can* say with more confidence, however, is that exceptionally few barns, granaries, watermills or grain ovens – certainly none on a scale that would suggest substantial crop surpluses – have been found at the early Anglo-Saxon settlements which, from the fifth century onwards, constitute a new and distinctive type of settlement unlike any late Roman precedent (Hamerow

2012, 1). The implication is that, whatever arable practices may have lingered beyond *c*. AD 410 at 'very late' Romano-British sites, they were not typical in early Anglo-Saxon rural communities.

This is not to suggest that crops were unimportant at the earliest Anglo-Saxon farms. Rather, crop-handling activities such as storing, drying, malting or milling grain seem to have been undertaken on a sufficiently small scale that special facilities were not required. Distinctive structures for the storage of cereal grain (or any other crop) are notably scarce among excavated rural settlements throughout much of the Anglo-Saxon period, and indeed continue to be relatively rare in the medieval period (Gardiner 2011, 207). In their absence, Hamerow suggests, it is likely that 'unthreshed grain was stored in the rafters of houses, in *Grubenhäuser* ... and/or in ricks ... and then threshed as the need arose' (Hamerow 2012, 50). By implication, the appearance of specially-built storage structures at a site could indicate that harvests had become too large, or their handling too centralised, to be accommodated by these less formal means.

Archaeological identification of grain stores depends upon an understanding of the principles of grain storage. Until cooked or milled, harvested grains remain biologically active and therefore measures must be taken, not only to keep out pests such rodents and insects, but also to inhibit respiration in the grains themselves. Otherwise they may germinate, producing warmth and humidity which could foster fungal, bacterial or insect infestation and thus spoilage (Sigaut 1988, 8–12). Grain which will be stored for extended periods is often therefore subject to some degree of atmospheric control to reduce warmth and moisture (and hence also it may be dried in an oven, as discussed below). So, for example, the air supply may be restricted, as in the hermetically sealed storage pits of Iron Age Britain; or conversely ventilation may be encouraged through open-air storage or else a purpose-built granary with ventilation holes. Granaries or ricks constructed on raised platforms can offer the dual advantages of rodent-deterrence, through the addition of 'rat guards' between the granary and its legs, and all-round ventilation to keep temperatures and humidity levels lower (Sigaut 1988, 8–12). Besides these general considerations, it should also be remembered that different crops have different storage qualities. Cereals can be either hulled or free-threshing, a distinction which I will consider in more detail later in this chapter. In brief, the grains of hulled varieties of wheat remain inside their protective spikelets even after threshing, and can be stored in this form to provide an additional, natural barrier to spoilage (Hillman 1981, 138). Free-threshing cereals, by contrast, lose this level of protection immediately upon threshing: only in their bulky unthreshed form, as loose ears or else bundled into sheaves, do they retain the physical protection of their chaff (Sigaut 1988, 13–14).

With these functional considerations in mind, I have located only two examples in the study regions of probable granaries or haylofts at Anglo-Saxon rural settlements – identified as such by their excavators because of their regular arrangements of postholes in the manner familiar from some Romano-British sites including Orton Hall Farm. Specifically, what we are looking for here is a regular layout of postholes, including one or more central posts, with the minimum number necessary to support the elevated storage unit without providing additional opportunities for rodent access. These criteria are met by the nine-post structure 'N' at Wicken Bonhunt (Essex), of eighth- to ninth-century date; and by Structure B 2734 at Yarnton (Oxfordshire), of seventh- to eighth-century date (Wade 1980, 98; Hey 2004, 69, 124–125). The Yarnton structure has more than 20 postholes which, in the excavators' preferred reconstruction, could constitute one six-post and two nine-post granaries (Figure 49). However, there are no plant remains directly associated with any of these features, so we can determine neither what crops might have been stored there, nor in what form they were stored. Indeed, different kinds of arable produce might have been stored in the same structures at different times: 'clean' processed grain, for instance, or unthreshed whole ears, or hay, pulses, flax seeds, or still other crops.

Another structure at Yarnton, of similar date, may also be relevant to this discussion. Structure B 3624 is a circular, concentric arrangement of postholes, a little over ten metres in diameter, with no associated finds. It is interpreted by the excavators as a fowl-house or dovecote, but this interpretation is entirely conjectural, and I believe that a closer parallel may be found among helms: open-air crop storage structures consisting of raised platforms upon which harvested crops are stacked or thatched (Figure 50).

Helms have been more often identified in German and Dutch archaeology but, as Gardiner demonstrates, there are possible examples from medieval England too, of which Yarnton's B 3624 and a smaller, near-contemporary structure at Raunds (Northamptonshire) are the earliest yet identified (Zimmermann 1992; Gardiner 2012, 27). In a helm, cereals, hay, or other suitable crops are stacked or thatched upon an outdoor elevated platform, sometimes beneath an adjustable roof, sometimes also around a central post, until removed for processing and consumption. Unlike enclosed granaries or barns, helms offer minimal shelter to insects or vermin, so their sheaves should in theory be less susceptible to pest damage. Such natural protection can be further enhanced by stacking the sheaves with the ears facing inwards and straw facing outwards (Sigaut 1988, 16–17; Zimmermann 1992; Gardiner 2012, 24–25). One potential disadvantage of this mode of storage is that, depending upon the precise manner of construction, it may be impossible to remove only a portion of the stack for threshing. Rather, once broken into, the whole store might have to be threshed and either consumed or removed to alternative storage facilities prior to further processing, thus demanding a considerable and rapid input of labour. Sigaut suspects that the stack originated as 'an occasional addition to the barn for exceptionally plentiful harvests' (Sigaut 1988, 16). Hence, the stack, rick and helm may plausibly be seen as innovations directly related to agrarian growth, requiring high labour investments and responding to the needs of increasing productivity.

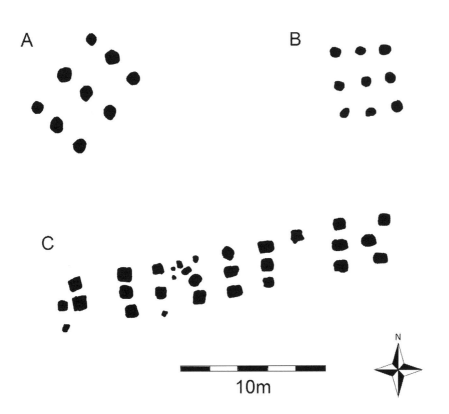

FIGURE 49. Plans of putative granaries at (A) Wicken Bonhunt (after Wade 1980, 97, fig. 38); (B) Orton Hall Farm (after Mackreth 1996, 90, fig. 57); (C) Yarnton (after Hey 2004, 127, fig. 6.23)

It has not been possible to identify any other closely comparable structures in the study regions although, considering their ephemeral nature, the apparent scarcity of helm-type structures in the archaeological record is hardly surprising. There is even some ambiguity in recognising Yarnton's candidate. A different arrangement of postholes (structure B 3619) is described by the excavators as rectangular or sub-rectangular, but it is sufficiently close to being circular for it to be mistaken for the 'fowl-house' (or helm) in Hamerow's reproduction of the site plan (Hamerow 2012, 97–98, fig. 3.18). The close proximity of both of these structures to rectangular post-built edifices perhaps recalls Sigaut's model, in which stacks are used to store the overflow from other buildings in the event of a bumper harvest (Figure 51)

Admittedly, however, there is nothing to suggest that these other rectangular buildings were barns. Following Sigaut, I take 'barn' to mean a large roofed building in which cereals were stored unthreshed, in the ear, often with the straw still attached and possibly bundled together into sheaves (Sigaut 1988, 16). Such buildings make no specific attempt to control the storage atmosphere of the crop. They may incorporate a threshing and winnowing floor, allowing sheaves to be brought down and processed as and when required. They are

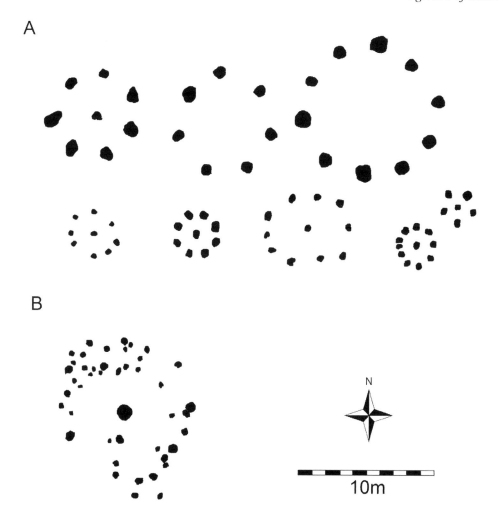

A

B

N

10m

FIGURE 50. Comparison of Yarnton structure B 3624 (after Hey 2004, 113, fig. 6.9) with helms identified in Germany and the Netherlands (after Zimmermann 1992, 36, fig. 2)

therefore well-suited to settlements producing cereal harvests so large that immediate processing of the whole volume would be impractical. Despite the lack of a diagnostic ground plan, barns may arguably be identified in excavated settlements on circumstantial grounds. Gardiner suggests two criteria which may be considered highly suggestive of barn-type functions: nearby concentrations of charred cereals, and internal post-holes which could indicate a raised floor for storage purposes, comparable to the central posts in the 'aisled barns' familiar from early medieval farms in the north-west of continental Europe (Gardiner 2012, 28–29; Hamerow 2002, 37–38). One structure at Yarnton, of eighth- to ninth-century date, might perhaps meet these criteria, but it would be the sole identified example in the study regions.

Hall B 3348 has modest indications of a central internal division (one or two posts), and the nearby enclosure ditches and associated pits produced dense

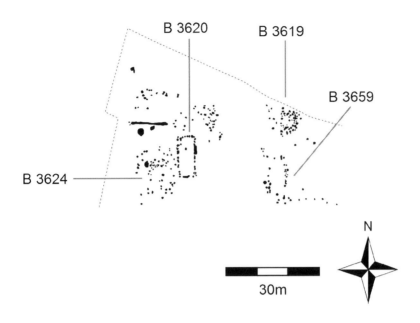

concentrations of charred grain, including a remarkably rich sample from pit 3693 containing a relatively high proportion of straw fragments, which may conceivably represent the burnt remains of stored sheaves. Hence, although Hall B 3348 lacks the long, narrow form of the continental barns, and is not assigned any special function by the excavator, there is circumstantial evidence to suggest that it could have served barn-like functions (Gardiner 2012, 29). It may also be significant that the building shares its ditched enclosure with a hearth (feature 3329) – potentially useful for drying crops – and fence lines which could have delineated zones for crop processing activities such as drying, threshing, winnowing and sieving (Figure 52).

It is interesting to note that the only other comparable 'barn' structures of like date, as identified by Gardiner, are at the putative royal tribute collection centre at Higham Ferrers (Northamptonshire, late seventh to early ninth century) and the estate centre at Bishopstone (East Sussex, eighth to tenth centuries) (Gardiner 2012, 29, citing Hardy *et al.* 2007 and Thomas 2010). To these might be added a 'barn and threshing floor' arrangement identified at Lyminge (Kent), from that royal site's eighth- to ninth-century monastic phase (Thomas 2009, 6–7).

It therefore seems plausible that barns were an innovative feature of some estate centres from around the eighth century onwards, sometimes – with hearths and threshing floors – forming part of larger crop processing units within zoned settlement plans. Yarnton's 'hearth', in this reading, takes the part of a grain oven, albeit on a more modest scale than the T-shaped, stone-built Roman examples discussed above. The term 'grain oven' is employed here as a synonym for the varied and largely interchangeable terms used in archaeological

FIGURE 52. Putative crop storage and processing unit at Yarnton (eighth to ninth century) (after Hey 2004, 154, fig. 13)

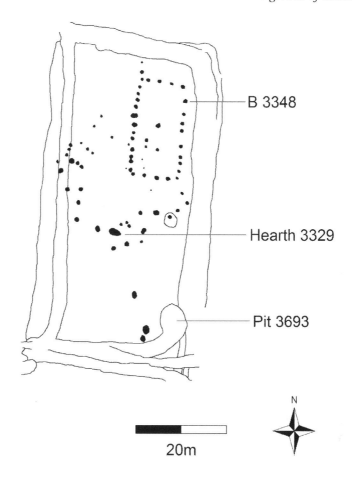

B 3348

Hearth 3329

Pit 3693

N

20m

literature, of which 'corn-dryer' and 'corn-drying kiln' are among the most common. In brief, a grain oven here denotes a structure which heats but does not cook cereal grains (other crops may be similarly processed, but current evidence suggests that cereals were the principal crop receiving such heat-treatment in this period). Heating in this way is intended to dry, parch or malt the crop. Drying may occur after the harvest, or after threshing but still before long-term storage, to ward against spoilage; or it may occur immediately before milling, to render grains harder and therefore easier to grind (Lenehan 1986, 1–8; van der Veen 1989, 303–305; Moffett 1994, 61). Parching is a subtly different function, specific to hulled cereals, intended to make their encasing chaff more brittle and thus easier to remove (Nesbitt & Samuel 1996, 42–48; see the discussion of cereals below). Finally, malting is the process whereby sprouting grains are 'cured' in hot air which ends germination without destroying the enzymes necessary for fermentation and the brewing of beer.

Specially-built grain ovens are not essential for any of these functions. We might rather imagine a wide spectrum of pre-industrial technologies, with the large purpose-built Roman-style affairs at one end and, at the other, the heated stones or round-bottomed pots observed by Fenton on modern Orkney and Shetland (Fenton 1978, 375). Yarnton's hearth, in my interpretation, would fall somewhere in between these two extremes: a small vernacular grain oven, informal, relatively ephemeral, but nonetheless purpose-built with the heating area separated from the stoking area by a short connecting flue, to ensure a gentle heat rather than direct toasting. The crop would be set upon a drying or malting floor, or 'kiln hair', above the heating area (Rickett 1975, 1–2, 19–28).

There are few or no well-dated grain ovens in the study regions, vernacular or otherwise, that appear to predate the seventh century AD. There are three possible ovens, of the small and vernacular kind, at Brettenham (Norfolk) and four or five at Brandon Road North in Thetford (Norfolk). All of these have been assigned a fifth- to sixth-century date, largely on stratigraphic grounds. But there are problems of residuality and intrusion at both sites, and it would be possible to argue instead for a late Roman date for these features (Mudd 2002, 67–68; Atkins & Connor 2010, 29). The same is true of the clay-lined feature 3405 at Handford Road, Ipswich, a considerably larger oven with a long, curving flue, for which the excavator considered a Roman or early Anglo-Saxon date possible (Boulter 2005, 26–27). More ovens can be assigned, generally with more confidence, to the seventh, eighth and ninth centuries, although the interpretation of these features specifically as *grain* ovens is in most cases conjectural. Some of these examples are small (fewer than 2 m in any dimension), such as the aforementioned Yarnton hearth, and others at Great Wakering (Essex), Stansted Carpark (Essex), Ely (Cambridgeshire) and Sedgeford (Norfolk) (Dale *et al.* 2010, 209–210, 219; Cooke *et al.* 2008, 183–184; Kenney 2002, 13–14; Davies 2008, 91). Larger examples exceeding 2 m in length are reported from Thetford, Sedgeford and North Elmham Park (Norfolk), Eynsham (Oxfordshire), Wolverton Turn (Buckinghamshire) and, again, Great Wakering (Essex) (Atkins & Connor 2010, 33; Bates 1991, 5; Hardy *et al.* 2003, 42; Wade-Martins 1980, 69–73; Preston 2007, 95; Dale *et al.* 2010, 209). Larger still – over 4 m in length, and rivalling Romano-British examples in their size and masonry – are the ovens situated just outside the study regions at Higham Ferrers (Northamptonshire), Feltham (Middlesex) and Gillingham (Dorset) (McKerracher 2014b), all probably dateable to the 'long eighth century' period. Whatever their economic function, these abnormally large ovens may also have served as status symbols, deliberate stone statements of *Romanitas* by classicising landlords. Discussing the villa-owning élites of classical antiquity, Purcell has argued that their investments in facilities such as wine and oil presses were not only of financial benefit, but also bore aesthetic appeal: that the *villa rustica* could be a cultural expression as well as an economic unit. Perhaps a similar phenomenon could be proposed in the context of early medieval lordship (Purcell 1995, 157–173; McKerracher 2014b).

It is likely that Anglo-Saxon grain ovens, of any type, are under-represented in the archaeological record because of difficulties in both dating and discovering such features. As Hamerow suggests, they might often have been situated well away from settlement cores in order to reduce fire hazards. Perhaps this is why the ovens at Stansted and Higham Ferrers, for instance, appear relatively isolated in their landscapes (Hamerow 2012, 155). Nevertheless, the most convincing examples that I have identified in the study regions are of seventh-century or later date. Like the barns and granaries, they appear to be an innovation of the long eighth century, following a post-Roman hiatus. The fact that several of my proposed examples are not especially large does not necessarily rule out an association with large crop surpluses. They may still have been sufficient for the piecemeal processing of a substantial harvest in relatively small batches, as and when required. It may even have been deemed a more efficient use of labour to spread out the tasks of crop processing over the wintertime, whereas immediate, large-scale processing in bulk at harvest time would have required a more intensive input of labour and fuel over a shorter period. This latter approach, one could conjecture, might have been more practicable on a Roman villa estate with a large body of servile labour. Such a strategy of prolonged wintertime processing might even be hinted at in the estate management tract known as the *Gerefa*, a literary work of *c.* AD 1100, which lists threshing (*ðerhsan*) and building 'a stove at the threshing-floor (*on odene cylne macian*) – for an oven (*ofn*) and kiln (*aste*) …' among the reeve's duties in winter (Swanton 1975, 26).

Watermills, by contrast, are a technology with altogether different implications. They represent a much greater initial investment, both in the structure itself and in water-management expertise, but they arguably also offer a more constant return, reducing the repetitive labours of manual milling. Again, it is possible that they served to some extent as status symbols, but their existence generally suggests that good harvests were reliably expected at a site – harvests big enough to justify mechanized milling. We have already considered the possibility that some Roman mills (such as at Ickham, Kent) continued operating into the fifth century. But there is no evidence for the construction of new mills until the later seventh century and thereafter. Among the most elaborate, if poorly published, are the remains of a triple vertical watermill and leat at Old Windsor (Berkshire), believed to be a royal site (Wilson & Hurst 1958, 184). Dendrochronology gives a probable construction date for the Old Windsor complex in the late seventh century, making it broadly contemporary with the late seventh- or early eighth-century watermills discovered at Barking (Greater London), Northfleet (Kent), Wareham (Dorset), and Wellington (Herefordshire) – no two of which are quite alike – and a millstone fragment of similar date excavated at Dorney (Buckinghamshire) (Fletcher & Tapper 1984, 119–120; Denison 2001; Blair 2005, 256; Tyers in Andrews, Biddulph, *et al.* 2011, 92–93; Roe in Foreman *et al.* 2002, 37). A channel to the south of the excavated settlement at Wicken Bonhunt, presumably belonging to that site's eighth- to ninth-century occupation phase, has been speculatively interpreted as a leat for a watermill (Wade 1980, 96). While it is entirely conjectural, there being no associated structural remains to account for the mill itself, this interpretation would at least be consistent with the national chronological pattern.

Preservation conditions could potentially exercise a strong bias against the survival of watermill remains which, as watercourses change over time, are likely to become vulnerable to riverine erosion, especially those mills smaller than the elaborate complex at Old Windsor (Watts 2002, 17–18; Wilson 2007). The archaeological scarcity of Anglo-Saxon watermills could therefore be misleading, as it certainly is for the eleventh century. The Domesday Survey of 1086 records thousands of mills, including a proliferation in the study regions, of which practically none are known archaeologically (Watts 2002, 83).

Nevertheless, limited though it is, current excavated evidence does seem to indicate a hiatus in watermill construction between the fifth and late seventh centuries, with something of a revival, however modest, thereafter. This chronology echoes those outlined above for granaries, barns and grain ovens, a synchrony which plausibly suggests some common cause. As I have already proposed, investment in special structures for storing and processing crops is unnecessary unless volumes have become too large to handle by less formal, more manual, *ad hoc* means. This would imply either that harvests became more abundant around the late seventh century, or that their storage and processing became more centralized or nucleated at this time. Perhaps both of these things occurred. Direct evidence from the arable environment may help us to understand this issue better.

Arable environments

The dynamic nature of a ploughed field makes it a difficult object of study for the archaeologist (see Chapter 2). Cultivation leaves other traces in the earth, however, which are potentially more durable and dateable than the fields themselves, namely pollen and sediments. Pollen grains are reproductive cells which travel within and between plants to enable fertilization. They are microscopic and often air-borne. Their nature and analytical value are therefore distinct from those of plant macro-remains such as charred seeds, which are visible to the naked eye (Cappers & Neef 2012, 132–134). Where pollen has been able to accumulate in undisturbed layers of sediment over time, a selective stratigraphic record of local vegetation builds up, which the palynologist can sample at intervals to assess ecological changes over time. Unlike (most) plant macro-remains, pollen sequences need not be dependent on past human activity for their creation and preservation. They are therefore sometimes able to elucidate the wider vegetation history of an area beyond the immediate environs of an excavated farm (Dark 2000, 2).

Some pollen sequences from the study regions indicate little or no vegetation change over the period in question. At Sidlings Copse (Oxfordshire), for instance, open grassland seems to have prevailed throughout both the Roman and the Anglo-Saxon periods (Day 1991, 467). There are several other sequences, however, in which an increased percentage of cereal pollen has been taken to signify an increased emphasis on crop husbandry – an expansion of cultivated land, say – from the seventh century onwards, within and beyond the study regions (Rippon 2010, 58–59; Hamerow 2012, 147). This inference of arable expansion from increasing proportions of cereal pollen is not as straightforward as it may sound. With the exception of rye, cereals release little pollen to the air and, owing to the relatively large size of the pollen grains, that which is released travels only short distances. Cereals may therefore be systematically under-represented in pollen cores unless grown or processed in the immediate vicinity of a catchment, and unimpeded by taller vegetation (Wiltshire 1990, 19; Dark 2000, 5; Wiltshire in Williams & Newman 2006, 124–125; Hale in Ivens *et al.* 1995, 416–417). Arable expansion is a plausible but not exclusive interpretation of increasing percentages of cereal within a pollen sequence. Such fluctuations could reflect changes in the location rather than the extent of cereal cultivation (i.e. indicating that cultivation drew closer to the pollen catchment); and it is possible in some instances that cereal pollen was deposited in the dung of livestock, potentially leaving a false arable signature on pastureland (Greig in Hey 2004, 377).

Nevertheless, there is an intriguing, compelling consistency to the chronologies of those sequences exhibiting a proportional increase in cereal pollen in the Anglo-Saxon period. And the argument that these rising percentages of cereal pollen do in fact reflect an expansion in arable farming is strengthened by the recognition of an increased rate of sedimentation, not only in the same period, but often in the very same deposits that produced the pollen sequences. Such acceleration in sedimentation (or alluviation) is frequently and plausibly interpreted as the result of increased local erosion, itself a likely result of expanding cultivation: more soil being ploughed, and fewer roots remaining to hold that soil together, combining to cause more sediment to be washed into the low-lying catchments.

What patterns do the pollen and alluvial sequences present in the study regions (Figure 53)? At Brandon, Suffolk, an increase in both cereal pollen percentages and sedimentation follows a local conflagration that occurred in the early- or mid-seventh century, and seems to accompany the foundation of the (probably monastic) settlement nearby (Wiltshire 1990, 12–15). At Micklemere, Suffolk, a comparable trend in both pollen and sediment data has been radiocarbon-dated to between the seventh and tenth centuries (Wiltshire 1988; Murphy 1994, 29–31). The same dual trend is seen again in the Oakley palaeochannel on the Norfolk/Suffolk border, where an earlier, slighter, proportional increase in cereal pollen is dated between AD 445 and cal. AD 670–820, with a later, more pronounced increase between cal. AD 670–820 and cal. AD 710–890 (Wiltshire in Ashwin & Tester 2014, 405–421). Increases in cereal pollen, potentially datable to the seventh century, are also seen at Hockham Mere, Norfolk, and Stansted, Essex (Sims 1978, 58; Wiltshire & Murphy in Havis & Brooks 2004, 353–354).

FIGURE 53. Sites with pollen or sedimentary evidence discussed in the text

A later development is suggested by the Easterton Brook palaeochannel near Market Lavington (Wiltshire) where percentages of cereal pollen do not increase until cal. AD 770–950 (Wiltshire in Williams & Newman 2006, 124–127). In the Upper Thames valley, acceleration in sedimentation also has a somewhat later chronology, occurring at the Drayton and Yarnton floodplains only from the ninth century onwards (Booth *et al.* 2007, 19–20). In the Nene valley too, between the two study regions, alluviation is said to have accelerated only from around the ninth century onwards (Brown & Foard 1998, 82). Hence, taken together, this palaeoenvironmental evidence could well suggest a date in the seventh century for the beginnings of arable expansion in East Anglia and Essex, and by the ninth century for areas further west. The pollen and sedimentary evidence suggests that the reappearance of granaries, mills and crop ovens from the seventh century onwards was a response, not only to a centralisation of crop storage and processing, but to a genuine expansion of crop production in the landscape.

The uneven distribution of all the palaeoenvironmental sequences, and the inevitably wide ranges of the radiocarbon dates involved, should caution against too rigid an interpretation of the geographical and chronological patterns. As already discussed, these data represent the incidental, 'natural' peripheries of human activity, not the activity itself. They constitute the penumbra of farming, along with the granaries, ovens and mills, which altogether point convincingly towards a growth in crop husbandry in the landscape, broadly dated between the seventh and ninth centuries, across the study regions. To come any closer to the heart of the process, to gain a firmer grip on arable growth in the long eighth century, we must turn to the very grains that were its basic product.

Introducing the charred plant remains

British archaeobotany relies, for the most part, on ancient misfortune. Whereas animal bones are the inevitable, constant by-product of pastoral farming, the fruits of crop husbandry are generally only bequeathed to posterity if some accident prevents them from being eaten, worn or otherwise consumed in antiquity. The most common means of preservation is the accidental burning of harvested crops, which under favourable conditions turns them into those staples of archaeobotany, charred plant remains – by some margin, the most widespread and abundant form of plant remains in British archaeology (Hall & Huntley 2007, 9–10). Such remains survive where grains, seeds and other items have burned incompletely, leaving 'fossils' in carbon which are then invulnerable to microbial decay, and which can be recovered from excavated soil samples by flotation in water. The lighter carbonized items float free of the heavier, sinking sediments. Charred plant remains – specifically, charred *crop* remains – are something of a paradox. Their creation, although unintentional, is a routine part of crop production. If the crop in question requires fire in its processing, as with cereals being dried, parched or malted, then there is a strong likelihood that some will periodically become charred. For this reason, charred plant remains, dating from between the fifth and ninth centuries, are not rare in the study regions. They are recorded at 96 sites, in 736 separate samples or deposits (Figure 54). Where a single context has been sampled more than once, I have treated the data for all sub-samples collectively, for analytical purposes. In this way, each 'sample' is taken to represent a (potentially) discrete episode of activity, an accidental artefact of arable farming (Jones 1991, 64; van der Veen 2007).

Many and various plant species are represented in this charred dataset, including cultivars such as peas, beans and flax, edible wild species such as black mustard and fat-hen, weeds such as corncockle and stinking chamomile, and woody perennials like hazel, brambles and elder. Unsurprisingly, however, it is the cereals which dominate the dataset. Cereals are present at all of the sites with charred plant remains, and in very nearly all of the samples (those which lack cereals are all too small to be significant, with fewer than thirty charred items in each). Moreover, individual samples tend to be characterised chiefly by cereals, and in particular by free-threshing cereals. This distinction is not gratuitous jargonising for the botanical cognoscenti. It is essential detail for uncovering the arable economies and ecologies of Anglo-Saxon England, for treating charred deposits as a proxy for surplus crop production. Before pursuing that line of enquiry, however, we must briefly digress to review what the distinctions between the different cereal types actually mean in practical terms.

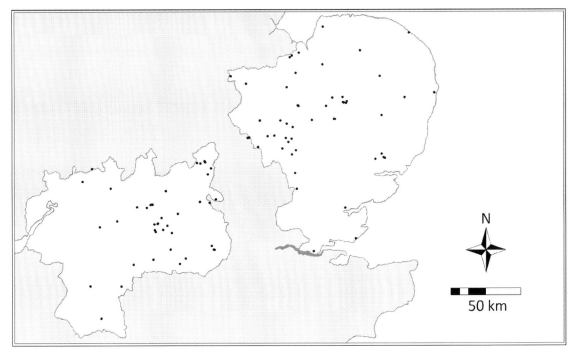

FIGURE 54. Distribution of sites with charred plant remains in the study regions

Cereals are cultivated and domesticated grasses, selectively 'bred' for characteristics such as hardiness, fruitfulness and ease of harvesting, to various degrees in different combinations (Cappers & Neef 2012, 380–387). The basic anatomy of the cereal plant, along with its terminology as used in this book, is illustrated in Figure 55. In brief, the straw (culm) is surmounted by the ear or *spike*, which comprises a spine of *rachis* segments, each of which supports a *spikelet*, which encloses one or more *florets* within two *glumes*. Each floret comprises a grain (caryopsis) held within a *lemma* and *palea* which may collectively be known as the lighter chaff, distinct from the heavier chaff consisting of rachis and glumes.

Broadly speaking, the cereals present in the project dataset can be divided into three categories. The free-threshing or 'naked' wheats release their grains from all surrounding chaff as soon as their ears are threshed. By contrast, the grains of hulled wheats (often called 'glume wheats' in archaeobotanical literature) remain tightly enclosed in their spikelets even after threshing, and require an additional stage called dehusking for the grains to be freed. To facilitate the dehusking process, the spikelets may first be parched to make the glumes more brittle, as discussed briefly above in the passage concerning grain ovens. Finally, between these two extremes, there are cereals such as hulled barley which are free-threshing in the sense that the heavier chaff (rachis and glumes) is lost upon threshing, but hulled in the sense that the lighter chaff (lemma and palea) requires an extra stage called hulling

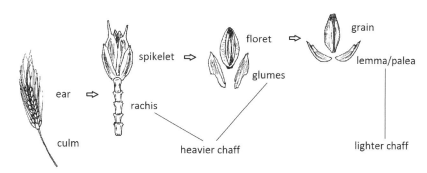

FIGURE 55. Schematic anatomy of the cereal plant for archaeobotanical analytical purposes

for its removal. These differences are illustrated in Figure 56 (see Hillman 1981, 130–139, fig. 4–7). However, since the lighter chaff is so fragile that it almost never survives among Anglo-Saxon charred plant remains, it is sufficient to draw the broad distinction between the hulled wheats on the one hand, and the free-threshing cereals (including hulled barley and oat) on the other, since each of these categories follows a largely consistent processing sequence.

These sequences normally result in the hulled wheats being represented archaeologically by charred grains and glume bases, and the free-threshing cereals by charred grains and rachis segments. The chaff, grain and any accidentally harvested weed seeds are separated out from each other using sieves at later stages of crop processing. The basic processing sequence for cereal crops admits little variation. The different stages of threshing, winnowing and sieving are likely to be common to cereal farmers widely separated in time and space. On this basis, modern ethnographic studies of traditional farming practices by archaeobotanists Glynis Jones and Gordon Hillman have identified key stages in that sequence at which plant items are likely to be preserved by charring, and they have demonstrated how the different stages can be distinguished by analysing the contents of archaeobotanical samples (Figure 57; Hillman 1981, 126–138; Jones 1984, 46). In the project dataset of charred plant remains, the great majority of samples are dominated by cereal grain and weed seeds in varying ratios, with very little chaff by comparison. These samples are therefore likely to represent the later stages of crop processing. This is precisely what we might expect of a dataset chiefly characterised by free-threshing cereals. The rachis and straw of these cereals are removed at an early stage, often traditionally outdoors, and so removed from settlement contexts where they might be accidentally burned and so preserved (Boardman & Jones 1990, 5–6; Jones 1987, 321–322). Incidentally, this is also why we should not read too much into a lack of chaff at individual sites, which need not mean that those settlements were 'consumer' sites receiving clean grain from elsewhere. Since practically all sites of this period produce little or no chaff, such an interpretation would misleadingly paint Anglo-Saxon England as an early consumer society.

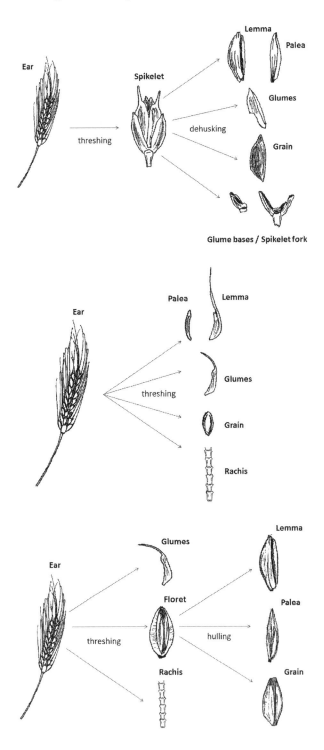

FIGURE 56. Simplified explanatory diagram of the processing of (A) hulled wheats; (B) free-threshing wheats; (C) hulled barley

Charred crop deposits and arable growth

So far I have described how charred plant remains are a widespread form of evidence in the study regions, how they are mainly characterised by cereals and associated arable weeds, and how such deposits are usually interpretable as accidental by-products from certain stages of cereal processing and storage. The argument that charred crop (and weed) deposits, although usually accidental, result from routine arable processing activities, is a crucial factor here. It implies a strong if probabilistic relationship between crop processing and charred deposits. An increase in the volume or frequency of cereal processing activities should, *a priori*, increase the probability of charring accidents and, by extension, increase the volume and frequency of charred plant remains entering the archaeological record. Van der Veen and Jones (2006, 222), considering Iron Age evidence from southern Britain, make the point succinctly: 'The answer to the question, "where are accidents involving

FIGURE 57. Highly simplified model of the cereal processing sequence, showing predicted products and by-products likely to be preserved by charring (after Jones 1984, 44, 55, figs 1 & 6)

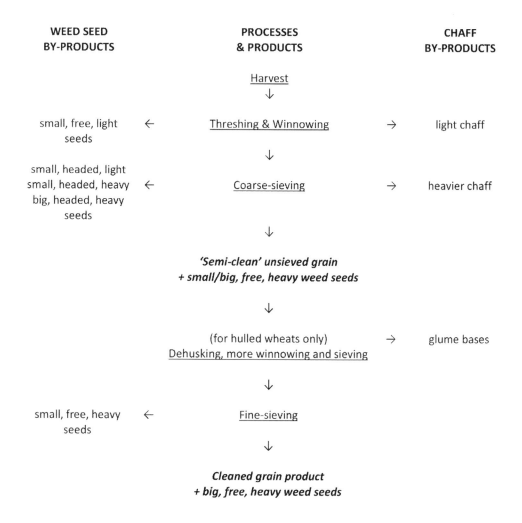

WEED SEED BY-PRODUCTS		PROCESSES & PRODUCTS		CHAFF BY-PRODUCTS
		Harvest ↓		
small, free, light seeds	←	Threshing & Winnowing	→	light chaff
		↓		
small, headed, light small, headed, heavy big, headed, heavy seeds	←	Coarse-sieving	→	heavier chaff
		↓		
		'Semi-clean' unsieved grain *+ small/big, free, heavy weed seeds*		
		↓		
		(for hulled wheats only) Dehusking, more winnowing and sieving	→	glume bases
		↓		
small, free, heavy seeds	←	Fine-sieving	→	
		↓		
		Cleaned grain product *+ big, free, heavy weed seeds*		

parching, drying and storage most likely to occur?" is that they will tend to occur in places where these activities are regularly carried out, i.e. where grain is handled in bulk.'

It would therefore be entirely logical to expect arable growth in the long eighth century to be manifested in a proliferation and growing abundance of charred plant remains: greater cereal destruction occurring in the wake of greater cereal production. There is, however, an important caveat. We first need to rule out the possibility that an apparent growth in the number or richness of charred botanical deposits is due simply to more rigorous and extensive sampling strategies. This potential bias can be mitigated in two ways. First, following the general philosophy of this project, we can focus on wider patterns across time and space, consistent patterns which transcend sampling and preservation biases at individual sites. Second, as with the animal bone assemblages discussed in Chapter 3, we can apply numerical filters to tune out the 'background noise' of very small samples, samples which may be too poor in charred plant items to be considered in any way representative of crop processing activities. The selection of a quorum can only ever be arbitrary, a balancing of analytical rigour against practical utility. For this dataset, I have found a minimum of 30 whole charred items per sample to be effective.

So does the charred dataset exhibit any signs of growth over time? Certainly the quantities of available evidence increase: both the number of sites with quorate samples and the total number of quorate samples soar in the eighth and ninth centuries (Figure 58 and Figure 59). The geographical distribution of these sites is shown in Figure 60, and again broken down by period in Figure 61. If charred plant remains are taken as a proxy for surplus arable production, then these data not only corroborate the palynological evidence in demonstrating growth in the long eighth century, but also fill out the picture by indicating that this phenomenon was not restricted to East Anglia. However, one problem with interpreting the samples in this way is that, as a general rule, largely due to pottery and coinage but also building traditions, evidence tends to become more identifiable and dateable from the seventh and eighth centuries onwards. Put simply, we stand a greater chance of finding and dating material of seventh-century or later date, regardless of its genuine abundance in the soil.

What this bias cannot manufacture, however, is artificially rich, dense deposits of charred plant remains. No matter how dateable the context or how zealous the soil sampling strategy, archaeologists cannot excavate really dense, grain- or seed-rich samples if the material was never deposited in such abundance in antiquity. Density, meaning the number of charred plant items per litre of sediment processed, is a key concept here. In itself, it can serve as a proxy for deposition rates. Rapid deposition of charred plant material results in a dense sample, whereas prolonged, gradual or piecemeal accumulation results in a sparse sample (van der Veen 2007, 987, table 6). The calculation of average density also calibrates for differences in sampling volume. A sample of 40 litres is theoretically likely to produce more charred plant remains than a sample of ten litres, but there is no reason why, in principle, it should contain more items *per litre* – unless it does indeed represent a genuinely richer, denser deposit.

In Figure 62, I have plotted the density of charred plant remains in all quorate samples for which relevant data were available. They are grouped by period, and presented in ascending order. There is a very clear trend towards increasing density over time, specifically increasing around the eighth and ninth centuries (Figure 62). Indeed, I have needed to omit two exceptionally dense samples of eighth- to ninth-century date from the graph, so as not to obscure the underlying patterns: one from Brandon, Suffolk, and one from Yarnton, Oxfordshire, each with more than 1000 items per litre. The denser samples of the long eighth century, with at least thirty items per litre, are not restricted to a single locale, although they do display the usual bias towards the Thames valley and Fen basin (Figure 63). Only three earlier samples stand out for their density of charred contents (greater than or equal to 60 items per litre), and these are either chronologically dubious, or exceptional for other reasons: one from Wilton (Wiltshire), which on stratigraphic grounds could well be Saxo-Norman (Pelling in De'Athe 2008, 21–27); and one each from Brandon and Alchester (Oxfordshire) which, although dateable to the fifth or sixth century on artefactual grounds, are both anomalously dominated by spelt wheat (Booth *et al.* 2001, 202–207; Murphy & Fryer 2009, 11). As will be discussed in Chapter 5, spelt was the favoured wheat of the Iron Age and Roman periods, but appears to have become exceptionally rare in the Anglo-Saxon centuries. Perhaps in these two samples, if their fifth- to sixth-century dates are accepted, we see another manifestation of that obscure, late twilight

FIGURE 58. Chronological distribution of sites with quorate samples

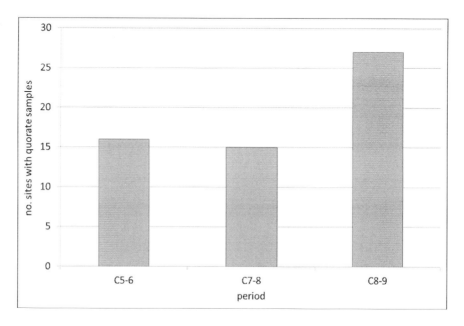

of Romano-British farming, the same phenomenon embodied by Alchester's corn-dryer which apparently continued in use into the fifth century (see above). For that mode of agriculture which we may more strictly term 'Anglo-Saxon', dense concentrations of charred plant remains seem genuinely to be a new phenomenon of the eighth and ninth centuries, the dark fruits of a process of arable growth which probably began in the seventh century. If the earliest Anglo-Saxon farmers were producing, handling and accidentally burning a large amount of grain, we have yet to find a trace of it.

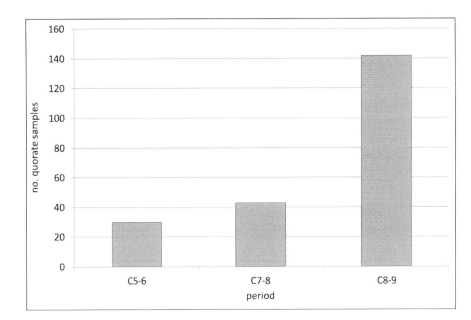

FIGURE 59. Chronological distribution of quorate samples

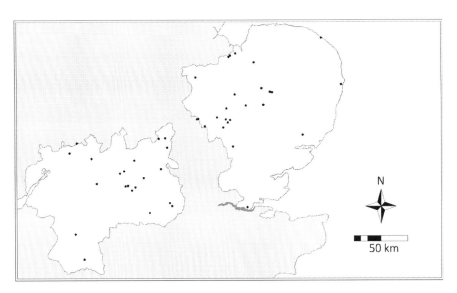

FIGURE 60. Geographical distribution of sites with quorate samples

FIGURE 61. Geographical distribution of sites with quorate samples in (A) the fifth to sixth centuries; (B) the seventh to eighth centuries; (C) the eighth to ninth centuries

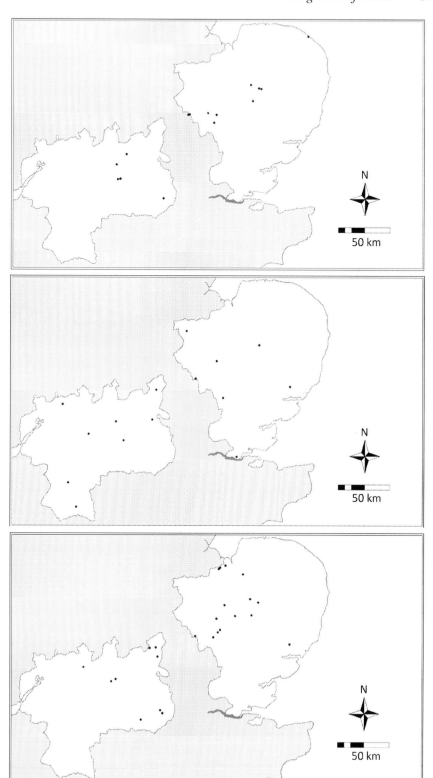

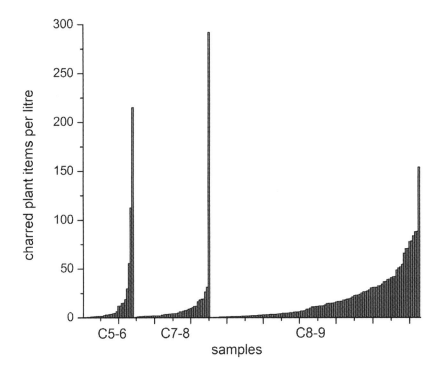

FIGURE 62. Average density of charred plant remains in quorate samples, grouped chronologically, excluding two exceptionally dense samples (see text)

FIGURE 63. Geographical distribution of sites with seventh- to ninth-century samples with at least 30 charred items per litre

Conclusions

In this chapter I have drawn upon several independent but complementary strands of evidence. All bear witness to the same general trend. A Romano-British farming regime which produced and processed substantial cereal surpluses persisted in some places beyond AD 410, but not everywhere, and probably for no longer than a century at the most. Otherwise, the scale of cereal production and processing in Anglo-Saxon farming remained relatively low until the seventh century, when arable growth seems to have begun. This growth continued, perhaps accelerating, through the eighth and ninth centuries. Evidence for this expansion spans the study regions, suggesting a widespread phenomenon that was not restricted to a single locale. A slight East Anglian bias in some strands of evidence, such as pollen, could reflect an overall bias in the archaeological record, or could conceivably reflect the environmental suitability of the region for cultivation: its favourable rainfall patterns, for instance (Williamson 2003, 35).

In any case, the evidence surveyed in this chapter is fully consistent with, and supportive of, the narrative which has driven the preceding chapters. I propose that arable farming expanded from the seventh century onwards, leading to greater harvests and hence more mills, granaries, ovens and charred plant remains in the archaeological record. As arable land grew, so pressure on hitherto marginal grazing land increased, and with it the need to exert greater control over livestock. With seasonal transhumance restricted, and the need to protect crops and homes from ranging livestock becoming ever more pressing, the same period witnessed growth in the use of hay meadows, droveways, and paddocks or stockyards. A parallel trend, afforded by this new pastoral infrastructure, saw cattle and especially sheep kept to greater ages in this period, thus extending their provision of secondary products such as traction and wool, respectively.

However, while I have cited a range of evidence that may help to demonstrate arable growth in the long eighth century, I have offered no proper description, let alone an explanation, of the process. In the next chapter, I will address this subject with extended reference to the archaeobotanical evidence. In this way, I will investigate how arable growth may have been achieved in terms of crops grown, environments exploited, and strategies pursued.

CHAPTER 5

The changing harvest

In terms of exciting modern interest, the ancient Roman diet enjoys more success than does the Anglo-Saxon. Potage, bread and ale lack the exotic appeal of stuffed dormice and rotting fish guts, and seem to have made little impression on the public imagination beyond the apocryphal burnt buns of King Alfred. 'Monotony,' writes one authority on Old English cuisine, 'may have been a characteristic of the Anglo-Saxon diet' (Hagen 2006, 392). The arguments presented so far in this book have done little to improve the dull reputation of the Saxon larder. Innovations such as specialised wool production and the growth of arable seem to offer little more to the feast than stringy old mutton and ever more potage. However, this is not the whole picture. It is the contention of this chapter that Anglo-Saxon harvests did not simply grow in the seventh to ninth centuries. They also changed, specialising or diversifying in response to various pressures including local environmental conditions, and so increased not only in quantity but also in quality – that is, in cultural and economic value. Crucially, further changes are also evident in the 'accidental harvest' of arable weeds, whose surviving seeds shed a unique light on the cultivation strategies and environments which form the ecological context of arable growth in Anglo-Saxon England. Thus, marshalling a range of archaeobotanical and palynological data, this chapter explores the changing harvests of cereals, other crops, and arable weeds between the fifth and ninth centuries.

We have seen already, in Chapter 4, how British archaeobotany is overwhelmingly dominated by charred plant remains, and how those charred plant remains are chiefly characterised by cereals. Indeed, it is possible to interpret many charred deposits as the burnt products or by-products of routine, everyday cereal processing. The project dataset of charred plant remains is therefore apt for elucidating patterns in cereal crop choices across time and space, for shedding light on the relative importance of different crop taxa in different regions and periods. Such investigations face the same profound methodological questions as this book's earlier investigations into livestock, and the relative importance of cattle, sheep and pigs. What do we mean by 'importance' in these cases? How can we discover the relationship between the bioarchaeological record and the original biological populations of antiquity? Although animal bones and charred plant remains are, in terms of composition and taphonomy, very different kinds of evidence, similar caveats apply to both. We have at our disposal only dead proxies, not authentic field observations. Hence, as in

Chapter 3, I will here focus upon regional and chronological differences in the data, upon patterns which, because sustained across several sites and samples, should transcend local biases in preservation, recovery and identification; and which, because strictly comparative, accommodate the different biases inherent in different species. Some species and plant-parts survive charring better than others, but these discrepancies should be common to the entire dataset, and therefore not responsible for spurious trends (Boardman & Jones 1990).

By 'importance', I mean agricultural precedence rather than, say, dietary contribution or cultural preference. Thus a crop which became more important is one from which larger harvests were sought and achieved, rather than one which necessarily fetched a higher price or made a more popular foodstuff. The greater the relative importance of a crop, the more frequently and abundantly it was sown, grown and reaped. How can this attribute be measured and compared archaeobotanically? Two approaches are taken here. First, I have undertaken presence analyses of the relevant plant taxa. Presence analysis is a semi-quantitative method which calculates the percentage of sites or samples in which a given taxon has been identified, for a particular region or period. This approach offers a broad perspective on vegetation history, by characterising general patterns in the occurrence of different taxa (Hubbard 1980, 52–53). While this method is ideal for viewing the general shape of the dataset and its taxonomic composition, it has the significant drawback of being a binary system which masks the complexity of archaeobotanical data. It distinguishes between presence and absence, but not between the abundant and the negligible, the concentrated and the sparse, and any shades in between. Presence analyses can exaggerate small differences in the archaeobotanical record, and diminish substantial variations. 'Presence' could equally denote one seed or a thousand seeds (G. Jones 1991, 64–65).

For these reasons I employ another, complementary approach alongside presence analysis: fully quantitative analysis of the different plant taxa in each sample. This follows the same basic method applied to zooarchaeological assemblages in Chapter 3, by calculating the percentages represented by the taxa in a sample. However, the application of this method to archaeobotanical data is rendered more complex by the greater diversity and taphonomic intricacy of charred plant remains. Indeed, Hubbard rejected fully quantitative botanical analyses on the grounds that the differential biases introduced through crop processing and charring, for example, opened too wide a gulf between a charred deposit and its 'parent economy' (Hubbard 1980, 51). These objections can only be overcome by rigorously standardising the data in question, so that – as far as possible, in terms of plant parts, species and agricultural activities – we are comparing like with like (G. Jones 1991, 64, 69). As with the animal bones, I have also applied an arbitrary but practicable quorum. Unless otherwise stated, the percentages of different taxa in a sample have only been calculated where the total sum of items belonging to those taxa is at least 30.

Wheat, barley, oat and rye

Such is the necessary methodological preamble. What cereal species are present in the charred dataset? A great many charred cereal remains are indeterminate, but the majority of grains have been identified to genus or species level. The most common genera represented are barley (*Hordeum* L.) and wheat (*Triticum* L.), with oat (*Avena* L.) and rye (*Secale* L.) less widespread but not insignificant (Figure 64).

Barley exists in numerous cultivated varieties – dense and lax-eared, hulled and naked, six-row and two-row – but charred grains and chaff are seldom distinctive enough to allow the archaeobotanist to make such close identifications (Moffett 2011, 351). While many and various barley names are used in the original reports from which the project dataset is drawn, overall there is little evidence to contradict Moffett's assertion (*ibid.*) that most barley in Anglo-Saxon England was hulled six-row barley (*Hordeum vulgare* L.), a free-threshing cereal. Among the wheat remains, indeterminate grains are widespread; they are not sufficiently distinctive to allow the archaeobotanist to identify them more precisely than '*Triticum*'. Nearly as common, however, are the free-threshing wheat remains, which in the original reports are often glossed as, or likened to, bread wheat (*Triticum aestivum* L.). The names '*Triticum aestivum sensu lato*' and '*Triticum aestivo-compactum*' are often seen, but the latter is not legitimate nomenclature,

FIGURE 64. Presence analysis of cereal types, among 705 samples and 96 sites with any charred cereal remains

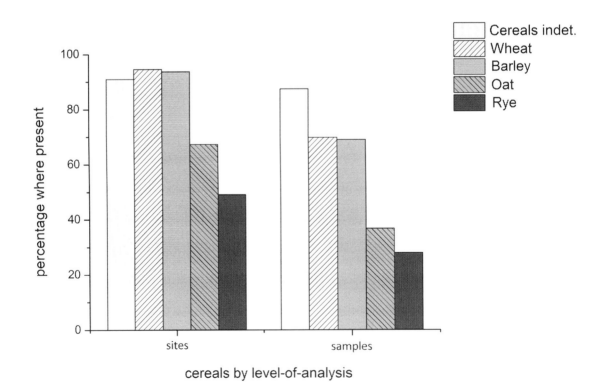

cereals by level-of-analysis

and in any case charred free-threshing wheat grains are notoriously difficult to identify to species, while the more diagnostic chaff is rare in charred deposits (M. Robinson pers. comm.; Moffett 1991, 233–235). Ultimately, the question of which free-threshing wheat species were cultivated in Anglo-Saxon England need not detain us here. Archaeobotany is not botany, and Linnaean binomial precision is not always necessary or appropriate. For the purposes of this study, it is enough to group all such identifications under the umbrella category of free-threshing wheat (*Triticum* L. free-threshing).

Rather less common are the hulled wheats emmer (*Triticum dicoccum* Schübl) and spelt (*Triticum spelta* L.), plus some occasional, ambiguous items which have not been identified more exactly than emmer/spelt (*T. dicoccum/spelta*). While I sometimes group all such items collectively under the umbrella term of hulled wheat (*Triticum* L. hulled), the species-level distinctions are important in the analyses which follow so I have preserved them in the dataset where possible.

Rye (*Secale cereale* L.) is the least common of the cereals in the charred dataset, but arguably the simplest to describe since it entails no species-level ambiguities; I treat *Secale cereale* as a single, consistent taxonomic category. Oat (*Avena* L.) is more problematic because wild and cultivated strains may both be present, but their grains are indistinguishable. Floret bases (part of the chaff) can be more diagnostic, but these are rarely preserved and still not a truly definitive guide to distinguishing the wild from the cultivated (Jacomet 2006, 52). The majority of oat remains in the charred dataset are indeterminate grains which can only be classed generically as *Avena*, potentially a crop plant but potentially an arable weed, or even something in between: a wild but tolerated (because edible) contaminant growing amongst cultivated crops. Given that oat grains are far more common than most wild seeds in the dataset, sometimes even dominating samples, I have generally given *Avena* the benefit of the doubt and erred on the side of 'potential crop' rather than 'potential weed'. Both rye and oat are free-threshing cereals, because they lose their heavy chaff upon threshing.

As described in Chapter 4, the charred dataset is chiefly characterised by cereals; the cereals are dominated (where discernible) by free-threshing species; and the samples are dominated (where discernible) by clean sieved grain, or sieved-out weed seeds, or the unseparated mix of grain and weed seeds prior to sieving. Chaff of any sort is scarce by comparison. With this material at our disposal, and with the basic methods outlined above, we can address two key research questions pertaining to the Anglo-Saxon cereal harvest: when did free-threshing wheats replace spelt as the favoured wheat crop of post-Roman Britain, and when was hulled barley surpassed by free-threshing wheat as the most favoured cereal crop of all in early medieval England? The oft-presumed ascendancy of free-threshing wheat between the fifth and ninth centuries is a strong theme in studies of Anglo-Saxon farming, and therefore of vital importance in this book (Banham 2010, 179; Moffett 2011, 348–351; Hamerow 2012, 146). Given that free-threshing wheat, especially bread wheat, is thought to thrive particularly well on rich clay soils, could its ascendancy be linked directly to the expansion of agriculture into clayey environments from the seventh century onwards (see Chapter 2)?

It has long been accepted that, whereas spelt was the predominant wheat crop of Roman Britain, as evinced by many archaeobotanical data, free-threshing wheat came to replace it in the Anglo-Saxon period (Green 1981, 133). The chronology, speed and nature of this fundamental shift in staple crops remain somewhat obscure, however, and likewise the reasons behind it. It is usually supposed, more or less implicitly, that the demise of spelt was a rapid process that occurred fairly promptly after the withdrawal of Roman governance, during the fifth or (at the latest) sixth century. Hence, spelt remains found in Saxon contexts are often taken to represent residual prehistoric or Roman material, not authentically early medieval plant remains (Pelling 2003, 103). This assumption applies to hulled wheats generally, both spelt and emmer: if identified in post-Roman contexts, they are looked upon with suspicion. This orthodox perspective has, however, been challenged. Emmer seems only ever to have been a minor crop in Roman Britain but, so Pelling and Robinson have argued, it may have been reintroduced between the fifth and ninth centuries, perhaps locally to the Thames valley as part of an imported agrarian tradition of the Anglo-Saxon settlers (Pelling & Robinson 2000). Their key evidence comprises emmer glume bases radiocarbon-dated to cal. AD 435–663 and 670–900, discovered at Dorney (Buckinghamshire) and Yarnton (Oxfordshire) respectively.

Presence analyses of the charred dataset suggest, on first inspection, that hulled wheats *per se* may not be as scarce in post-Roman contexts as is usually thought (Figure 65 and Figure 66). In the fifth and sixth centuries, indeed, hulled wheats exhibit marginally higher presence values than do free-threshing wheats. The ascendancy of free-threshing wheats at the expense of hulled wheats is evident only from the seventh century onwards. Examining the different hulled wheat species in closer detail, we can trace a net decline in the presence of spelt over time, and a small rise in the presence of emmer, although spelt maintains the higher presence values overall throughout the period (Figure 67 and Figure 68). The presence analyses therefore suggest that the transition from spelt to free-threshing wheat was rather later and/or more protracted than is often assumed to be the case, with the free-threshing varieties achieving clear dominance only from the seventh century onwards. Over the same period there appears to have been a small rise in the presence of emmer, broadly in keeping with Pelling and Robinson's reintroduction model, but it remains a persistently rare crop, even by comparison with the dwindling spelt.

The fully quantitative analyses, on the other hand, clearly show how the presence analyses have exaggerated the significance of hulled wheat macrofossils which, in most samples throughout the fifth to ninth centuries, occur in very low quantities relative to free-threshing wheat remains (Figure 69). The proportions of spelt and emmer in most samples are arguably too low to indicate deliberate cultivation and processing, but may rather represent 'volunteer'

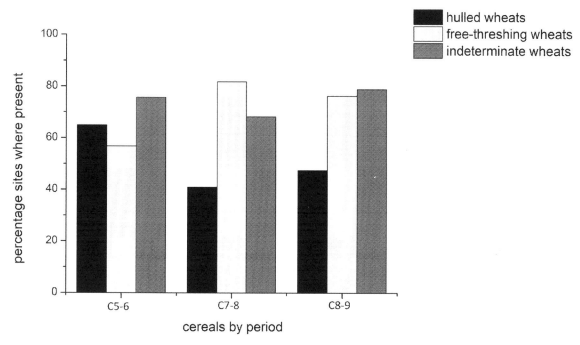

FIGURE 65. Chronological presence analysis of wheats among sites with charred cereal remains (total sites: 37 in C5–6, 22 in C7–8, 38 in C8–9)

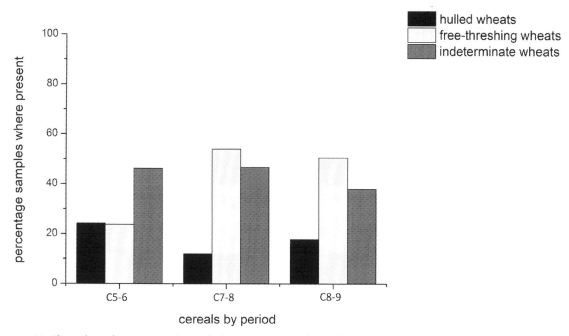

FIGURE 66. Chronological presence analysis of wheats among samples with charred cereal remains (total samples: 190 in C5–6, 150 in C7–8, 309 in C8–9)

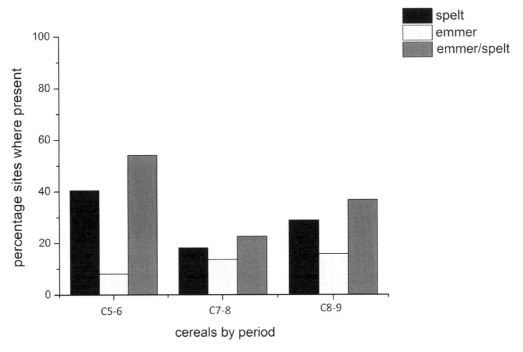

FIGURE 67. Chronological presence analysis of hulled wheats among sites with charred cereal remains (total sites: 37 in C5–6, 22 in C7–8, 38 in C8–9)

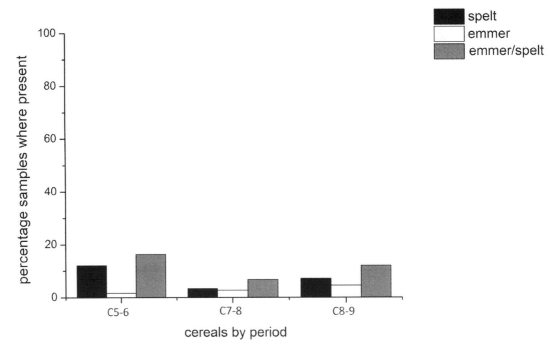

FIGURE 68. Chronological presence analysis of hulled wheats among samples with charred cereal remains (total samples: 190 in C5–6, 150 in C7–8, 309 in C8–9)

crops (essentially self-perpetuating contaminants) persisting in former spelt fields long since turned over to free-threshing cereals (Murphy 1994, 37). Alternatively, these traces of hulled wheat could represent residual prehistoric or Roman material, as is so often supposed. A different explanation is required, however, for those few samples in which hulled wheats constitute an unusually high proportion of the wheat remains – high enough to suggest, with some plausibility, that they represent the fruits of deliberate hulled wheat cultivation.

Hulled wheats predominate over free-threshing varieties in only 16 out of 57 samples (Figure 69). Three of these samples – one each from Alchester (Oxfordshire) and Brandon and West Stow (Suffolk) – can be dated to the fifth or sixth century, and may represent the lingering traces of Late Roman spelt cultivation beyond AD 410. Such relics would be analogous to the Alchester grain oven, for instance, which continued in use into the post-Roman period (see Chapter 4). The remaining 13 samples, however, are mostly dateable to the eighth and ninth centuries and are concentrated at four sites: Yarnton (Oxfordshire) and Dorney (Buckinghamshire), where emmer prevails; and Harston Mill (Cambridgeshire) and Thetford (Norfolk), where the abundant hulled wheat is spelt. In addition, although compatible data were not available for inclusion in the project dataset, concentrations of eighth- to ninth-century spelt have also been noted at Great Wakering (Essex) and Gloucester (Fryer in Dale *et al.* 2010, 219; Green in Heighway *et al.* 1979, 186).

Emmer may well represent a localised innovation at Yarnton and Dorney, as Pelling and Robinson have proposed. As for the spelt, it is surely straining credibility to suggest that late Roman spelt cultivation 'lingered' for as long as 400 years at a few settlements which in other respects appear to be culturally Anglo-Saxon. A more credible hypothesis, I would argue, is that hulled wheats continued to offer some advantage which was lacking in free-threshing wheats, which encouraged the active renewal or reintroduction of these varieties in the long eighth century. One potential advantage afforded by emmer and spelt is their greater natural resistance to spoilage. They can be threshed and winnowed and yet their grains, unlike those of free-threshing wheats, remain enclosed in their protective spikelets. Hence, hulled wheats may have been more desirable than their free-threshing counterparts if a crop was required for longer-term storage and transportation – as would have been the case in Roman Britain, for example, with the transfer of corn surpluses between villas, towns, garrisons and markets.

With the resurgence of trade, urbanism and crystallising lordship in the long eighth century, not to mention a renewed interest in the classical Roman tradition, it is conceivable that hulled wheats once again came to occupy a vital niche in Anglo-Saxon agriculture as the portable, storable crop of choice. Consider the nature of the sites with marked concentrations of hulled wheat remains. Dorney is thought to represent a rural market, Great Wakering an ecclesiastical centre, Yarnton and Harston Mill farms with trade links (witness their Ipswich Ware), Gloucester a minster town and Thetford destined to

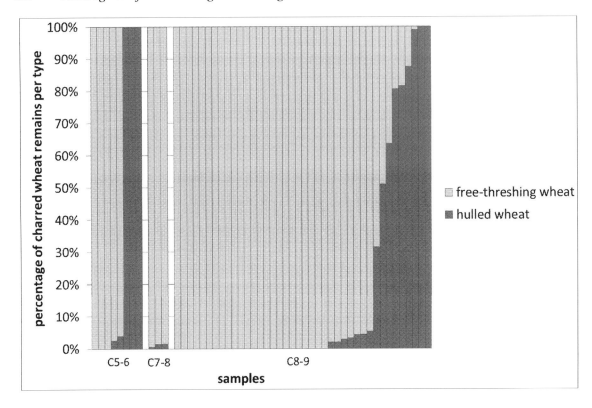

FIGURE 69. Graph of samples, by period, showing percentage of hulled wheats relative to free-threshing wheats

become one. Anglo-Saxon emmer and spelt may possibly be more widespread and abundant than we currently appreciate since, in the absence of indisputably early medieval radiocarbon dates or stratigraphy, a sample dominated by hulled wheats will generally be dated (almost by definition) to a prehistoric or Roman phase. The argument for the immediate post-Roman demise of spelt wheat risks becoming circular.

Having considered the evidence of the charred dataset, therefore, I would offer a provisional, speculative revision of the traditional narrative regarding Anglo-Saxon wheat cultivation. In the fifth and sixth centuries, spelt farming dwindled at all but a few sites, mirroring the fate of other aspects of late Roman agriculture such as grain ovens and watermills. By the eighth and ninth centuries, the changed socio-economic context had created a new niche for spelt – and emmer too, in some parts – to occupy, as wheats better suited to transportation, exchange, and longer-term storage.

None of this speculation alters the fact that the charred dataset, and Anglo-Saxon archaeobotany in general, is overwhelmingly characterised by free-threshing cereals: barley, wheat, rye and oat. The relationship between these four free-threshing cereal taxa is a crucial matter to be considered. If there is an orthodox view on this subject, it is that hulled barley was the most important cereal crop between the fifth and seventh centuries but that free-threshing wheat rose to first importance thereafter. The origins of this model stretch back as far as 1944 (see McKerracher 2016b, 89 on the development of this 'bread wheat thesis'), but it has been developed most fully by Banham, who cites archaeobotanical data as well as documentary references to the lower status of barley relative to wheat. Indeed, she has argued that an Anglo-Saxon cultural preference for the bread made from free-threshing wheat was ultimately responsible, not only for a change in the precedence of crops, but also for the adoption of mouldboard ploughs to till the damp earth favoured by wheat, and hence also the creation of ridged-and-furrowed open fields (Banham 2010, 182–187). While Banham also posits an increase over time in the importance of rye and oat in Anglo-Saxon England, it is the shifting balance between hulled barley and free-threshing wheat which lies at the core of her hypothesis. As discussed below, I believe that this project's charred dataset reveals a quite different pattern to that presented by Banham. However, I concur with her view that crop choices have a wider significance tied into the development of agricultural strategies (see, generally, McKerracher 2016b on this question).

In the analyses which follow, I have grouped indeterminate wheat remains with those of free-threshing wheat in those instances (the majority) where few or no hulled wheat remains have been identified. This not only follows the logic of proportional reassignment (i.e. sharing out indeterminate counts between other taxa in proportion to their known relative abundance), but also recognises the fact that free-threshing wheat grains are more prone than those of hulled wheats to distortion in the charring process, and thus more likely to be deemed indeterminate during analysis (Boardman & Jones 1990, 8).

The presence analyses, as illustrated in Figure 70 and Figure 71, do not consistently support the model of barley losing precedence to wheat from the seventh or eighth century onwards. The pattern among *sites* approximates to Banham's proposed trend (Figure 70), but that among *samples* does not (Figure 71). In any case, it is highly questionable as to whether the differences between the presence values for wheat and barley in the site-based analysis are truly significant. They are arguably small enough to be explained by accidents of preservation, recovery or identification, rather than representing real historical trends. Both wheat and barley are practically ubiquitous crops throughout the regions and periods under study, according to the presence analysis data, with little perceptible expansion or contraction over time. By contrast, Banham's other observation – that rye and oats became more important over time – is more strongly and consistently borne out by the charred dataset. The spread of rye from the seventh century onwards is particularly pronounced, and a somewhat earlier development than that envisaged by Banham (Figure 70 and Figure 71).

Much the same can be said of the fully quantitative results. For clarity, data for each cereal have been graphed separately (Figure 72 to Figure 75), but the percentages relate directly to each other, i.e. the same samples and calculations underlie all four graphs. In this way, it can be seen that the proportions of wheat and barley grain do not change dramatically over time, although there

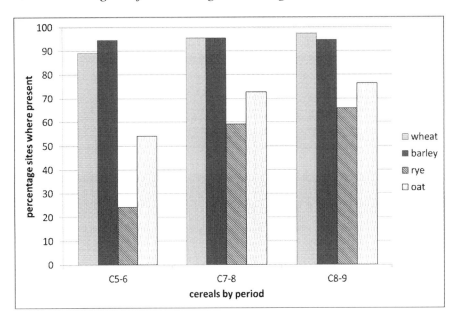

FIGURE 70. Site-based presence analysis of wheat, barley, oats and rye, by period (total sites: 37 in C5–6, 22 in C7–8, 38 in C8–9)

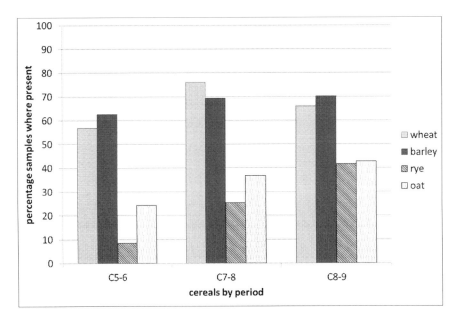

FIGURE 71. Sample-based presence analysis of wheat, barley, oats and rye, by period (total samples: 190 in C5–6, 150 in C7–8, 309 in C8–9)

is perhaps something of a polarisation from the seventh century onwards: the percentages of each crop cover a greater range in the later samples than in the earlier samples (Figure 72 and Figure 73). More striking is the appearance of samples which are comparatively rich in rye and oat as a new phenomenon of the seventh century and later (Figure 74 and Figure 75). What, if anything, might these patterns have to do with the contemporary growth in arable surpluses posited in Chapter 4?

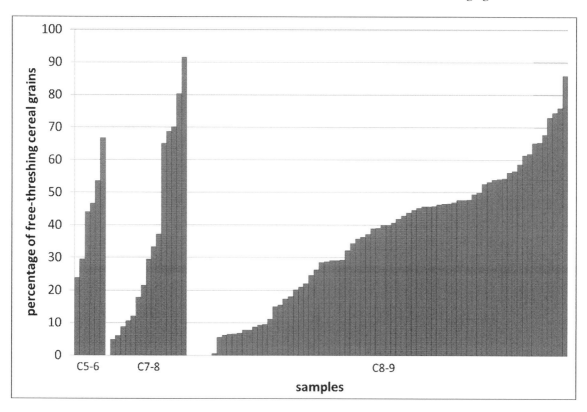

The answer may lie in the geographical dimension to the crop data. If we map the data from the quantitative analyses as interpolated maps, following the same methodology as used for cattle, sheep and pig percentages in Chapter 3, then a compelling pattern emerges. The evidence, which is largely of seventh-century or later date, shows clear correlations between crops and environment. Thus barley has concentrations in the Middle Thames valley, the Suffolk Coast, Breckland and East Anglian Heights – all of which bear the lighter, poorer soils which barley tolerates well – and also in the silt fens, whose high salinity barley will again tolerate (Figure 76; Murphy 2010, 215; Moffett 2006, 48, table 4.3). Conversely, concentrations of free-threshing wheat – which is better suited to richer, clayey soils – are seen around the Midland Clays and edges of the peat fens, the southern East Anglian Plain, and the clay vales around the Upper Thames and Severn regions (Figure 77; M. Jones 1981, 107). Oat, which can tolerate very poor soils, has scattered localised concentrations, at Yarnton (Oxfordshire) and Ipswich (Suffolk) for example, but is most conspicuously concentrated at Chadwell St Mary in southern Essex, where the terrain is infertile, heavy and acid (Figure 78; Williamson 2003, 27). Finally rye, whose extensive root systems make it very drought-hardy, has marked concentrations in the sandy, droughty Breckland and the similarly dry and sandy Suffolk coast near Ipswich (Figure 79; Moffett 2006, 48, table 4.3).

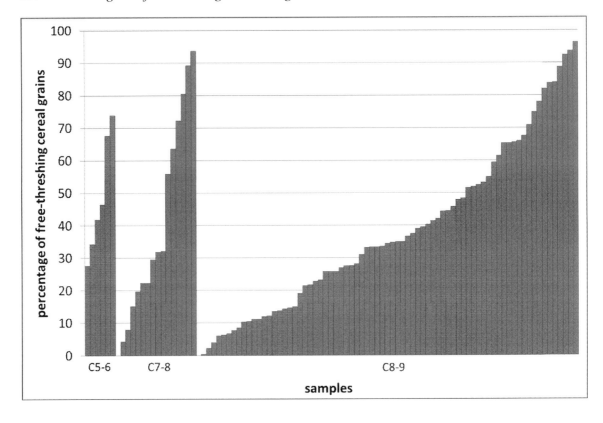

FIGURE 73. Percentage of barley grains in samples over time

It may not be very surprising, at least to the ecologically-minded observer, to find that certain cereal crops are most concentrated around the terrains to which they are biologically suited. But these patterns help to explain how the increase in arable surpluses discussed in Chapter 4 was achieved. From the seventh century onwards, I propose, cereal surpluses were increased not only by the expansion of free-threshing wheat cultivation onto hitherto marginal interfluves among the fertile clay vales, but also by 'fine-tuning' other crop choices in response to local environmental conditions, whether moist and fertile or dry, poor and acidic. The rise of rye and oat in the long eighth century shows that the growth of arable was not all about free-threshing wheat and heavy clays.

Yet it must be admitted that this model is partly circumstantial. I have alluded to the prevailing soil conditions in different regions in very general terms, overlooking the subtle variability of local environments on the ground. In Chapter 1, I emphasised how most sites in the dataset are well situated to exploit two or more kinds of landscape, whether gravel terraces, chalk uplands, clay plateaux or sandy heaths. How can we be sure that a particular crop was being grown on a particular terrain at a given site, without simply assuming automatic local correlations between, say, rye and sand? Counts of grain and chaff cannot provide this confirmation, but the seeds of arable weeds preserved amongst the charred cereals have the potential to shed some light on the matter.

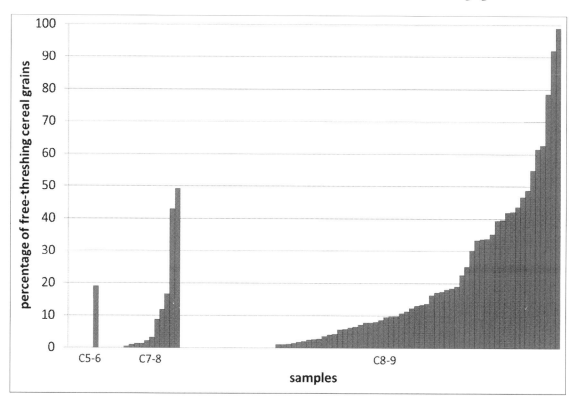

FIGURE 74. Percentage of rye grains in samples over time

The accidental harvest

'Weed', as a category of plant, is partly biological and partly cultural. It is a plant in the wrong place, such as a wildflower growing amongst cereal crops, and specifically one which benefits from its culturally inappropriate habitat (Mabey 2010, 5–14). So the wildflowers growing among cereals may benefit from anthropogenic enrichment of the topsoil (i.e. muckspreading); and, if their seeds are sufficiently grain-like to survive hidden amongst the seedcorn, humans even assist in their propagation, returning their seeds to the soil each year. Thus are the tares sown among the wheat. In an age before chemical herbicides, and in systems too extensive to ensure completely efficient removal of weeds from the field, it was inevitable that arable farmers would reap an accidental harvest of weeds. Some weed seeds can be removed along with the chaff through threshing, winnowing, coarse-sieving and fine-sieving. But others, especially the so-called 'crop mimics', remain concealed amidst the corn. Preserved in charred deposits alongside the grain and chaff, they offer to the archaeobotanist a unique window on lost arable environments. By comparing the weed floras preserved in different samples, and with reference to the known ecological preferences of the wild species represented, we can begin to discern patterns in the ancient growing environments of crops.

Given that the charred dataset contains more than 200 potential weed taxa, it is impractical to review presence data for all of these. So, in accordance with my

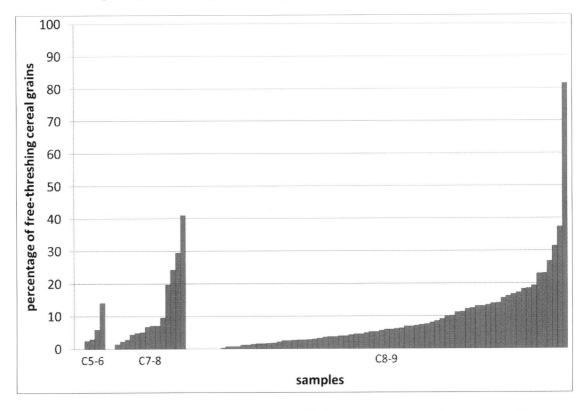

universal precept to focus on overall trends, I will highlight those species which display the greatest net change in presence values between the fifth and ninth centuries: a change of at least 15% in terms of sites, or 10% in terms of samples. As usual, these are arbitrary but practicable criteria. The results are shown in Figure 80 and Figure 81. In all cases, the diachronic change is an increase, i.e. all the species listed occur in a higher proportion of samples, and at a higher proportion of sites, in the eighth and ninth centuries than in the fifth and sixth centuries. While the species do not seem to constitute a single, coherent group in ecological terms, they point generally towards wet soils, heavy clays, and nitrogen-enriched ground. Spike rush grows in wet conditions; stinking chamomile is characteristic of heavy clay soils and has the long seed-dormancy period necessary to survive in heavy ploughing regimes; and several other listed species are relatively nitrophilous (nitrogen-loving), including common or spear-leaved orache, common mallow, fat hen, knotgrass and especially stinging nettle and henbane (Kay 1971, 625; M. Jones 2009; Williamson 2003, 121; Stevens in Hey 2004, 362; Stace 2010). These patterns would be consistent with the extension of arable land onto previously uncultivated heavy, damp clays by the eighth century, along with the associated introduction (or reintroduction) of heavy ploughing, and an intensification of soil-enrichment strategies such as manuring, middening, or folding, to keep the fields in good heart (Williamson 2003, 79–81).

FIGURE 75. Percentage of oat grains in samples over time

When we turn to individual samples and the quantities of weed seeds contained therein, our work is complicated not only by the difficulty of distinguishing true arable weeds from other wild taxa, but also by the sheer number and diversity of both species and samples in the charred dataset. One possible solution to this problem is to apply multivariate statistical methods to accommodate the high volume and complexity of the data. I discuss this approach and its results in detail elsewhere (McKerracher 2014a, 251–275). It is sufficient here to note the most prominent and significant trends. So stinking chamomile, for instance, the weed most strongly associated with heavy clays and heavy ploughing, is most prominently represented in two groups of samples. First, its seeds occur in relatively large numbers in those spelt-dominated samples at Alchester and Brandon which I interpret as relics from the fifth-century twilight of late Roman agriculture, which still used heavy ploughs and clay soils. When proportions of stinking chamomile seeds rise again, it tends to be in those eighth- to ninth-century samples from Yarnton (Oxfordshire), Ely (Cambridgeshire) and Gamlingay (Cambridgeshire) which are also relatively rich in free-threshing wheat grains. Thus weed ecology provides independent supporting evidence for a link between arable expansion onto heavy soils and free-threshing wheat cultivation in the long eighth century (as argued by Banham 2010).

By contrast, rye-rich samples from eighth- to ninth-century Ipswich are more distinguished by their seeds of corn spurrey (*Spergula arvensis* L.) and wild radish (*Raphanus raphanistrum* L.), both of which species are ecologically associated with acidic, sandy soils and other non-calcareous terrain (Stace 2010, 418, 467; Clapham *et al.* 1962, 133–134, 257–258). This is especially significant because it is not true of all samples from this phase of activity at Ipswich. Other samples, typically with a more even balance between the proportions of rye, oat, barley and wheat grains, have a less marked weed-ecological signature. It might be, therefore, that the emporium of Ipswich was receiving its staples from two different sources in its variegated hinterland: harvests dominated by rye from more specialist farming regimes along the sandy Suffolk coast, and less monotonous harvests from more diverse regimes inland on the East Anglian Plain. In each case, these hypothetical farming groups could have been following a strategy aimed at ensuring and enhancing surplus production: on the coast, the fine-tuning of crop choices to best suit local conditions; and on the variable, undulating southern East Anglian Plain, the spreading of risk across different crop species with different ecological preferences and tolerances, such that at least one should return a good harvest each year.

Certainly, much more comparative work remains to be done on Anglo-Saxon weed ecology. But my preliminary, tentative forays into the data have begun to reinforce the impressions gained from the charred crop remains: that the diversifying arable farmers of the long eighth century mixed and matched crops and terrains both to safeguard and to maximise surplus production. This process did include the extension of ploughland onto heavy clays, especially for

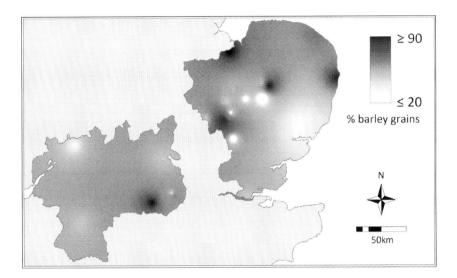

FIGURE 76. Interpolated map of percentage barley in samples

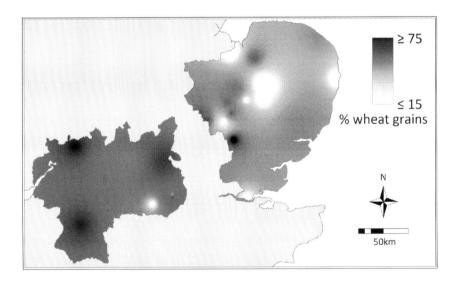

FIGURE 77. Interpolated map of percentage free-threshing wheat in samples

the cultivation of wheat, but that is only one part of a bigger picture of cereal farming in this period. And cereal farming itself must also be considered within a wider arable context, including a range of other crops.

Beyond the cereals

However important cereal crops were to the Anglo-Saxons, and still are to archaeobotanists, they were not the only crops growing in the farmed landscapes of early medieval England. The problem is that none of these

FIGURE 78. Interpolated map of percentage oat in samples

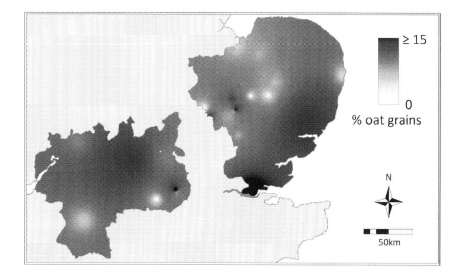

FIGURE 79. Interpolated map of percentage rye in samples

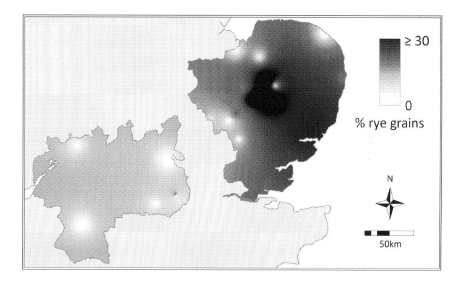

other crops is so prone to routine preservation by charring as are the cereals, because of their different processing requirements (Green 1982, 43). It is true that charred material does not account for the entire archaeobotanical record. Other, less cereal-oriented material does survive through waterlogging and, in deposits such as latrine pits and middens, by mineral-replacement. But there is not the systematic, predictable relationship between either of these modes of preservation and any crop processing sequence, such as there is between charring and cereal processing. Perhaps the closest analogy is the relationship between waterlogging and plant-fibre production. Flax (*Linum usitatissimum* L.) can be grown for its edible oily seeds, but also for linen production. In order

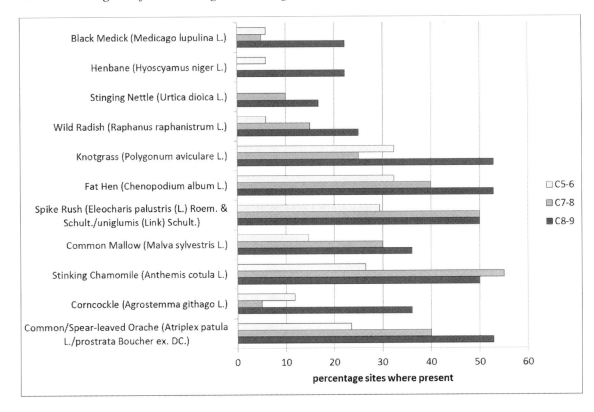

FIGURE 80. Selective weed presence analysis by sites (total sites: 34 in C5–6, 20 in C7–8, 36 in C8–9)

to extract the fibrous matter from its woody stem, one must soak flax in water, a lengthy and malodorous process known as retting (Pals & van Dierendonck 1988, 242, table 1). Prior to retting, flax is threshed ('rippled') to remove seeds and capsules, which might thus also become waterlogged if rippling took place near the pool or stream used for retting.

Waterlogged evidence for flax processing in the project dataset is very slim, but consistent with an expansion from around the seventh century onwards (Figure 82). The only earlier waterlogged deposit with appreciable quantities of flax remains is the fifth- to sixth-century fill of a well at Barton Court Farm (Oxfordshire). These seeds and capsule fragments could represent dumped rippling waste, although whether one considers this evidence to be early Anglo-Saxon or very late Romano-British is a moot point, given that the site shows little break in occupation between the fourth and fifth centuries (Robinson in Miles 1986). Certainly the evidence, both charred and waterlogged, is much more abundant for the seventh century and later. From the waterfront zone in the seventh- to ninth-century settlement at Brandon (Suffolk), for example, comes a charred sample with abundant flax seeds and capsule fragments identified as rippling waste burnt as fuel, along with waterlogged fibrous stem matter interpreted as 'scutching' waste, i.e. from the stage with follows retting (Fryer & Murphy in Tester *et al.* 2014, 325). In addition, other waterlogged

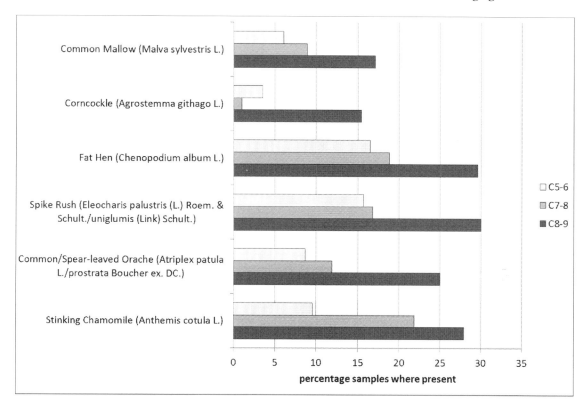

FIGURE 81. Selective weed presence analysis by samples (total samples: 115 in C5–6, 101 in C7–8, 240 in C8–9)

deposits from the same zone contained seeds of elder (*Sambucus nigra* L.) and weld (*Reseda luteola* L.), both of which species can be used to produce dyes. It is therefore plausible that Brandon's waterfront samples represent not only textile production but also dyeing activities (Fryer & Murphy in Tester *et al.* 2014, 329–330).

A well at Westbury-by-Shenley (Buckinghamshire), material from which produced radiocarbon dates in the range cal. AD 653–772, contained flax seeds, stem and capsule fragments, which have been interpreted as retting waste (Letts in Ivens *et al.* 1995, 419–422). At Yarnton (Oxfordshire) meanwhile, waterlogged seeds and capsule fragments from an eighth- to ninth-century well are interpreted as rippling waste; and a waterlogged flax beet (bundle), capsule fragments and seeds in the nearby Oxey Mead palaeochannel are thought to represent a retting site, radiocarbon-dated to cal. AD 660–890 (Robinson in Hey 2004, 367, 408). Finally, similar evidence for retting – none of which is necessarily any earlier than the seventh century – has been discovered at three sites in the St Aldate's area of Oxford, in what would have been a seasonal floodplain in the early medieval period (Brown in Durham 1977, 170–172; Robinson in Dodd 2003, 80, 370, 382). Between the two study regions at West Cotton (Raunds, Northamptonshire), flax seeds and capsules interpreted as retting waste returned radiocarbon determinations centred on the eighth century (Chapman 2010, 29)

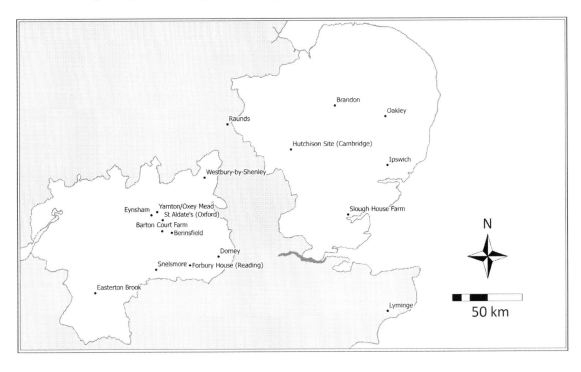

FIGURE 82. Map of sites mentioned in discussion of non-cereal crops

Available evidence is therefore consistent with a post-Roman decline, and a seventh- to ninth-century revival, of flax cultivation and linen production. A similar pattern of seventh-century growth can also be posited for hemp (*Cannabis sativa* L.), another fibre plant, and perhaps also stinging nettle (*Urtica dioica* L.), which often grows as a weed but can be harvested for fibre. A study of Danish retting pits from the Bronze Age to the Viking period found that these contexts frequently contain not only the seeds, capsules and stems of flax but also nettle seeds, hemp seeds and hemp pollen, raising the possibility that the pits were used for the retting of all three species (Troldtoft Andresen & Karg 2011, 520–523). *Cannabis*-type pollen (which could derive from either hemp or hops) occurs in sequences at Brandon (Suffolk), Oakley (Suffolk), Slough House Farm (Essex), Easterton Brook (Wiltshire), Snelsmore (Berkshire), and on the Yarnton (Oxfordshire) floodplain: none of these occurrences is demonstrably earlier than the seventh century (Wiltshire 1990, 14–15; Wallis & Waughman 1998, 186–187; Wiltshire in Williams & Newman 2006, 126; Waton 1982, 76; Greig in Hey 2004, 377; Ashwin & Tester 2014, 404–421). Indeed, the peat columns from Brandon's waterfront produced not only *Cannabis*-type pollen but also fragmentary fruits of hemp, and some samples from this zone contained abundant waterlogged nettle seeds (Fryer & Murphy in Tester *et al.* 2014, 313–330).

Altogether, therefore, evidence for the production and dyeing of plant fibres increases significantly from the seventh century onwards, and is geographically widespread within the study regions. The evidence is too slim for us to gauge the importance of fibre and dye plants within the wider agricultural economy, but it gives a clear indication of their increased importance between the seventh and ninth centuries.

In fact, in general terms, the whole archaeobotanical record of cultivars seems to diversify from the seventh century onwards. Grape (*Vitis vinifera* L.) pollen appears in the Oakley (Suffolk) palaeochannel sequence at some point between the seventh and ninth centuries, giving us the earliest known evidence for post-Roman viticulture, and in the Easterton Brook (Wiltshire) sequence by around the tenth century (Wiltshire in Ashwin & Tester 2014, 405–421; Wiltshire in Williams & Newman 2006, 129). This evidence for Anglo-Saxon viticulture raises the possibility that charred grape pips discovered at Yarnton and Eynsham (Oxfordshire) and Dorney (Buckinghamshire), in seventh- to ninth-century contexts, came from English-grown vines rather than representing imported raisins. Opium poppy (*Papaver somniferum* L.), which may grow wild but can also be cultivated for culinary or psychoactive applications, is scarce in the Anglo-Saxon archaeobotanical record but again restricted to deposits of seventh-century or later date. Waterlogged seeds of seventh- to ninth-century date are known from Brandon, charred seeds of like date from Yarnton, Dorney, Addenbrooke's Hospital (Cambridge) and beyond the study regions at Lyminge (Kent).

The evidence for garden pea (*Pisum sativum* L.) and field bean (*Vicia faba* L.) is more widespread but sporadic, usually a few charred pulses amongst cereal grain, although there is one rare instance of waterlogged bean chaff at Berinsfield (Oxfordshire), confirming that beans were indeed processed as a crop (Robinson in Lambrick 2010, table A18:4). No regional or chronological patterns are evident for these pulses, although rare finds of charred lentils (*Lens culinaris* Medik.) – one from Yarnton and two from Forbury House (Reading, Berkshire) – are both of seventh-century or later date. Hop (*Humulus lupulus* L.), best known as a flavouring agent in brewing, is represented in the project dataset only at Ipswich, by a few charred and mineral-replaced remains in eighth- to ninth-century contexts. It could have been grown locally or else imported to this mercantile settlement, likewise the non-native dill (*Anethum graveolens* L.) whose waterlogged seeds also occur at the site.

Overall, then, although the evidence for each species individually is very sparse, collectively the data reveal a diversification in (potential) crop taxa from the seventh century onwards. This archaeobotanical phenomenon could well reflect a real diversification in agriculture and horticulture around the long eighth century. It is impossible to estimate the importance of any of these plants relative to the cereals in Anglo-Saxon England, and difficult to do more than simply comment on their occurrence and distribution. But it is well worth remembering that the central narrative proposed in this book, focused primarily upon corn, sheep and cattle, might not tell a wholly representative story. Somewhere in archaeobotany's peripheral vision are diverse other stories which we are not yet able to tell in full, but we know that they are there to be discovered.

Conclusions

These five chapters have zoomed in ever closer, from a geological panorama of the study regions, to the farms and fields of the Anglo-Saxon countryside, to the animals that grazed there, and finally to the crops and weeds that grew there. The archaeobotanical and palynological evidence discussed in this chapter gives a crucial contribution to the whole, wider picture. All of the arguments and interpretations presented in the previous chapters have required that arable surpluses increased from the seventh century onwards. This chapter has argued that such growth was achieved through a combination of local specialisation and general diversification. As arable land expanded and the plough encroached upon land hitherto grazed, farmers increasingly adapted their cereal crop choices to maximise the productive potential of their fields, mindful of the possibilities and restrictions of their different crops and environments.

Yet at the same time, the revived cultivation of flax, grapevines and hulled wheats – and perhaps also other crops such as hemp – is difficult to ascribe to environmental considerations. Such trends hint rather at the influence of economic and cultural factors, such as a renewed interest in classical *Romanitas*, for instance, or the particular demands of monastic landlords for wine and liturgical vestments, or else the requirements of burgeoning long-distance trade. Anglo-Saxon agriculture in the long eighth century was rooted in natural ground, but its development was governed by a society whose needs were rapidly outgrowing the retrenched simplicity of post-Roman farming. In this way, with nature as the engine and culture as the tiller, a course was set for the transformation of Anglo-Saxon farming.

CHAPTER 6

Farming transformed

The lie of the land has changed since 1977, when Hunter Blair remarked upon the lack of evidence for agricultural development in Anglo-Saxon England. Forty years later, wherever we look – to livestock, to cereals, to stockyards, to the retting pool or the vineyard – it is difficult to ignore the evidence, from at least the eighth century onwards, suggesting notable innovations in comparison with the earliest Anglo-Saxon period. The archaeological data now abound, and the emergent patterns are striking. This period, this so-called long eighth century, saw farming transformed within and beyond Anglo-Saxon England. The lie of the land was changing, and early medieval England was poised for an era of unprecedented agricultural and demographic growth, halted only in the fourteenth century by the Black Death. The age of Offa and Charlemagne saw political and cultural landscapes across Europe and around the Mediterranean crystallise in new and distinctly post-Roman ways, a process underpinned and powered by agricultural development.

The transformation of farming in the long eighth century was a truly international phenomenon. Field surveys, geoarchaeology, palaeoenvironmental and bioarchaeological studies all bear witness to the expansion of cultivated land and the growth of arable surpluses during the long eighth century, not only in England but also in Ireland, the Frankish realm, the Germanic and Scandinavian lands, Byzantine Italy, and Al-Andalus, the Iberian peninsula under Arabic rule (McCormick *et al.* 2014, 24–50; Hamerow 2002, 139–147; Ruas 2005; Arthur *et al.* 2012; Puy & Balbo 2013). Unsurprisingly, farming changed in different ways in these very different places, with substantial and predictable variance between the temperate north-west and the Mediterranean zone. So, for example, rye's relative importance in northern France is not paralleled in the Mediterranean south, while the agrarian expansion in contemporary Italy was more focused upon grapes and olives than cereals (Ruas 2005, 407; Arthur *et al.* 2012, 446–451). Within the more corn-orientated north-west, local variability exists within a general uniformity of direction shown by a range of archaeological evidence. Much is reminiscent of the trends discussed in this book. For seventh- and eighth-century Ireland, for example, there is a general increase in the abundance of charred crop remains and in the construction of watermills and grain ovens; at the same time, crop choices diversified to include taxa such as flax and peas or beans, and oats became a more important cereal crop (McCormick *et al.* 2014, 24–50). In the Germanic north-west of continental Europe, meanwhile, there is evidence for increasing crop-storage capacities in this period, besides the laying-out of new field systems, the intensive practice of turf manuring, and the increased importance of oats and especially rye (Hamerow 2002, 134–147; Behre 1992, 148–150; Henning 2014). For Carolingian Francia, documentary sources record a proliferation of mills, breweries, bakeries and other productive facilities – especially at monasteries – between the seventh and ninth centuries (Lebecq 2000, 134). Grain ovens were sufficiently notable in ninth-century Francia to be included and annotated on the St Gall plan, an idealised diagram of a monastery, situated opposite a granary and labelled '*locus ad torrendas annonas*' (http://www.stgallplan.org, accessed April 2017). Again around the ninth century, zooarchaeological data from northern France point towards a growing emphasis on sheep husbandry (Crabtree 2010, 129).

Such is the wider context within which Anglo-Saxon farming must be considered. The English data furnish a case study in agricultural development in temperate Europe during the long eighth century, a widespread process which saw cereal surpluses grow, processing technologies proliferate, animal husbandry specialise, crop husbandry diversify, and oats and rye gain in importance. But the Anglo-Saxon example is also important – and locally varied – in its own right, enmeshed as it is within the specific social, political, economic and cultural milieu of the Old English kingdoms.

The seventh century witnessed profound changes in Anglo-Saxon society. From a turbulent and shifting political landscape emerged the kingdoms and aristocracies that came to dominate later Saxon England. Their establishment and consolidation were emphatically asserted in lavish burials and lofty halls, wherein gift-giving and feasting both rewarded and cemented the loyalty of royal retainers. By the later seventh century the major kingdoms were well established, their rulers issuing law codes, granting tracts of land, and gathering food-rents. Christianisation, religious and cultural, was also gathering pace. Anglo-Saxon élites drew themselves and their kingdoms increasingly within the ambit of western Christendom – both Celtic and Roman – in which scholarly connections and long-distance travel encouraged the exchange of materials, ideas and classical knowledge, both sacred and profane. With Christianity came Latinate literacy, and with literacy the documented, theoretically perpetual granting of bookland: land worthy of meticulous recording and productive long-term investment. From the *eorl* down to the *ceorl*, the relationship between man and land, between farmer and field, became closer, more intense and more stable. From the seventh century onwards, in their homes, boundaries, and even in their graves, Anglo-Saxons were digging themselves in as never before. Yet at the same time, economic horizons grew ever broader as long-distance trade and exchange intensified, fuelled by increasingly specialized and large-scale craft production. Coinage, metalwork, pottery, imported quernstones, inland markets, coastal and riverine emporia, and rural smelting sites all bear witness to the economic step-change.

Uniting and supporting these trends towards landedness and economic effervescence was the rising tide of agrarian growth. Like Britannia before it,

England was becoming more and more tightly yoked to the plough. Farming was perforce the ubiquitous constant of Anglo-Saxon life, and agricultural development was the ultimate catalyst for socio-economic change. Following the arguments put forward in this book, it is now possible to propose a cohesive, archaeologically-grounded model of agricultural development in seventh- to ninth-century England.

In the fifth and sixth centuries, two approaches to farming co-existed. One was the very latest phase of late Roman agriculture, the exact nature and extent of which remain doggedly obscure but whose broad outline we can begin to trace. This was a system which, even after the end of Roman governance, was capable of producing dense concentrations of charred spelt wheat and barley, which utilised watermills and large stone-built grain ovens, and which is otherwise manifest in evidence for granaries, flax cultivation, or animal bone assemblages strongly focused upon particular species, such as sheep in the Cotswolds or cattle in the Upper Thames valley. Had there been a major, abrupt dislocation in farming practices at the turn of the fifth century, we would hardly expect Romano-British field divisions to have survived into the medieval period, yet this too is in evidence.

Alongside this 'latest Roman' system, outlasting it and perhaps evolving out of it in some cases, there arose a way of farming commonly thought of as 'early Anglo-Saxon' because of its association with early Saxon material culture. This was a system principally characterised by simplicity. Arable farming produced only small surpluses. It focused overwhelmingly and fairly uniformly on only two crops, hulled barley and free-threshing wheat, both of more or less equal importance at most farms across the study regions. Arable soils tended to be light and easily worked, while animal husbandry was largely unspecialised and extensive, in the sense that herds were not closely managed in terms of spatial control, breeding or culling patterns. Such a system of unspecialised self-sufficiency seems entirely consistent with a socio-economic context in which goods-exchange was minimal and short-distance, and farms drifted and shifted within the landscape. This was mobile adaptability, not fixed intensity.

The seventh century heralded change. Where mobility had been the key to sufficiency, now stability was the key to growth. Arable land bearing cereals expanded during the seventh century in East Anglia and also, by the eighth and ninth centuries, further west along the Thames valley. Settlement distributions in some areas indicate a reoccupation in this period of terrains hitherto abandoned by the ploughman since the Roman period, such as heavy clays and fenland silts. While such land may have provided pasture for transhumant herding in the fifth and sixth centuries, only now was it returned to the plough. This model helps to explain not only the expansion of long-term settlement beyond the light soils, but also the resurgence of arable weeds such as stinking chamomile, preserved increasingly amidst the harvested grain. By implication, too, use of the mouldboard plough must also have begun to spread at this time, or else such heavy, damp soils could not have been effectively tilled.

Hulled barley and free-threshing wheat remained, jointly, the most important cereal crops overall, but with greater local variation in their relative importance, and alongside an ever greater role for rye and oats. All of these developments in the cereal spectrum embody a process of adaptive growth: focusing on the crop or set of crops which could best exploit the ecological potential of a given arable landscape, such as wheat in clay vales, rye on droughty sands, or barley on saline silts. In this way, arable innovation applied just as much on lighter soils as it did on the newly tilled heavier ones. Such an ecologically-minded approach, matching crops with the terrains to which they are well-suited, chimes well with the greater stability of settlements in this period. As the attachment between farmers and local environments became stronger and closer, so their agrarian practices needed to become more specifically attuned to them. The success of this system in returning bigger yields is manifest in the increase in charred cereal remains, granaries, grain ovens and watermills in the archaeological record from around the late seventh century onwards.

There is also evidence of a wider diversification in crop choices, whose influences may be more cultural than environmental. Thus spelt and emmer found a niche in the new agrarian regime, their greater natural resistance to spoilage perhaps recommending their cultivation for trade and tribute, for the markets and landlords characteristic of the long eighth century. Viticulture had been introduced (or reintroduced) to East Anglia, perhaps as early as the seventh century, and grapes were reaching the Upper and Middle Thames valley by the eighth century, perhaps in response to monastic demands for Eucharistic wine, but perhaps also because secular élites enjoyed a Dionysian draught as they revelled in revived *Romanitas*. The increased cultivation of flax for linen production from the seventh century onwards could likewise be related to monastic needs, i.e. for liturgical cloths, and indeed some of the strongest evidence for specialised textile production (flax, dye-plants and a wool flock) comes from the probably ecclesiastical settlement at Brandon.

Arable expansion acted in tension, however, with the changing course of animal husbandry. Spreading ploughland encroached upon grassland and wood-pasture, constricting the availability of winter grazing ground, while arable fields in their turn required protection from ranging livestock. The response was to create massive ditched systems of droveways, paddocks and stockyards in the vicinity of settlements to control and organise the herds of cattle and sheep; and to manage and mow hay meadows to provide winter fodder for the enclosed stock animals. The ability to overwinter livestock in this period was so effective, in fact, that more and more animals were kept alive each year, herd structures becoming – in East Anglia, at least – more mature. The result would have been an increased source of wool and dairy goods, but also of the crucial traction and manure that fed back into arable farming. In this way, settlement stability allowed both pastoral and arable farming to grow, by obliging farmers to work the land harder, more intensively, in return for greater productivity which in turn helped to sustain the whole system.

To argue that natural factors strongly influenced *how* agriculture developed is not to explain *why* agriculture developed between the seventh and ninth centuries. Environmental influences (including climate change) may have helped a farmer to increase his yields, but they did not oblige him to do so. Population growth is likely, *a priori*, to have been both a cause and an effect of agrarian growth: more mouths providing the incentive, more hands providing the means. The same can perhaps be said of this period's more specific increase in the non-agricultural population – traders, artisans, clerics, monks and nuns – whose very existence depended on the availability of other people's agrarian surpluses, and so ensured a stimulating demand for them. Yet it is difficult to adduce either of these circumstances as a prime mover for agricultural development, since the food must logically precede the eaters. That is to say, that the initial capacity to trigger agrarian growth must have hinged on some factor which did not itself, in the first instance, rely upon that growth.

This factor, or combination of factors, must also have been independent of specific locations, given how geographically widespread agrarian growth appears to have been around the long eighth century. In this historical context, the key factor, the prime mover, was surely kingdom formation. By this I mean not only the establishment of kings over Mercia, Northumbria, and so on, but also the wider processes of political consolidation from the royal hall down to local lords, who were becoming not just leaders of kin or war-bands but lords over land (Wickham 2009, 157–161). They were in some sense claiming the inheritance of the defunct Romano-British villa-dwelling class, hence perhaps their revival of Roman land-surveying traditions (Blair 2013b) and possible concordances between Roman and Anglo-Saxon masonry grain ovens (Figure 83; McKerracher 2014b). Some of the high-status settlement complexes of the late sixth or early seventh century may even, arguably, recall the courtyard villas of later Roman Britain. Might the derelict remains of the latter have in some measure inspired the former?

Just as the processes of kingdom formation occurred right across Anglo-Saxon England, so also is there evidence of agricultural development far beyond the case study regions considered in this book. There is also a persistent, marked association between such evidence and high social or ecclesiastical status. Anglo-Saxon watermills, for instance, are known from such dispersed locations as Wellington (Herefordshire), Ebbsfleet (Kent), Barking (Greater London), Worgret (Dorset), Corbridge (Northumberland) and Tamworth (Staffordshire). Where known, these sites appear to have been of royal (e.g. Tamworth) or ecclesiastical (e.g. Barking) status (Denison 2001; Blair 2005, 256; MacGowan 1996; Tyers in Andrews *et al.* 2011, 92–93; Rahtz & Meeson 1992; Snape 2003). Large grain ovens are widely scattered too in the seventh to ninth centuries, occurring at Feltham (Middlesex), Gillingham (Dorset), Higham Ferrers and Raunds Furnells (Northamptonshire), Hoddom (Dumfriesshire), Sherburn (Yorkshire), Ebbsfleet (Kent), Chalton Manor Farm (Hampshire) and Hereford. Again, so far as we can tell, the associations of these sites are lordly

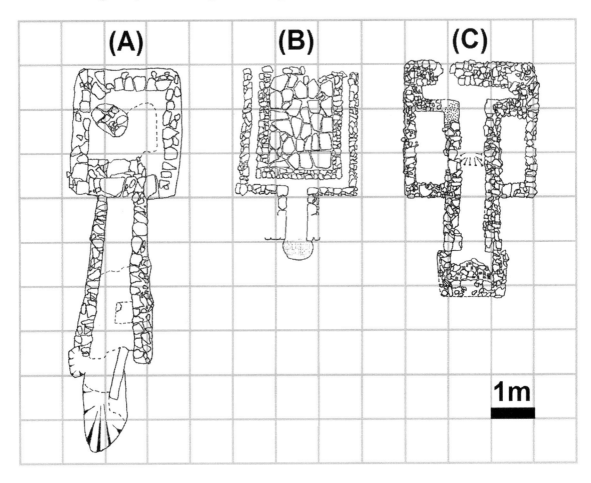

and/or ecclesiastical. Higham Ferrers is thought to have been a royal tribute-collection centre, Hoddom a monastery, Raunds Furnells the precursor to a manorial complex, Sherburn later became a royal vill, and Hereford was an episcopal seat (McKerracher 2014b; Holden in Lowe 2006, 100–106; Audouy & Chapman 2009, 66; Powlesland 2011, 6–7; Shoesmith 1982, 30–32; Andrews *et al.* 2011, 285; Hughes 1984, 72).

Turning to the environmental evidence for arable farming, sedimentary and palynological evidence in Devon indicate a marked increase in cultivation around the seventh and eighth centuries (Rippon *et al.* 2006). Rare post-Roman occurrences of spelt can be associated either with a long twilight of Late Roman farming – such as the remains at the former villa site at Northfleet (Kent), radiocarbon-dated to cal. AD 400–580 – or with high-status and ecclesiastical sites in the seventh to ninth centuries. The latter include Newton Bewley in the immediate hinterland of monastic Hartlepool (Co. Durham), the estate centre at Bishopstone (East Sussex) where spelt is radiocarbon-dated to cal. AD 687–895, or the eighth- to ninth-century phase at Lyminge, a royal

FIGURE 83. Comparative plans of: (A) a seventh-ninth century malting oven at Higham Ferrers, Northamptonshire (after Hardy *et al.* 2007, 52, fig. 3.33); (B) a Roman corn-drying oven at Rogging, Bayern, Germany (after Fischer 1990, 321, fig. 157); (C) a Late Roman corn-drying oven at Goring, Oxfordshire (after Allen *et al.* 1995, 41, fig. 30)

monastery (Smith & Stafford in Andrews *et al.* 2011, 304; Huntley & Rackham in Daniels 2007, 108–123; Ballantyne in Thomas 2010, 164–176; Campbell 2012). A possible association between spelt and status can even, perhaps, be traced in the Frankish evidence. Bakels has compared charred plant remains from excavated Carolingian sites in northern France with the documentary evidence of the *Brevium Exempla*, which contains a survey of royal estates written *c.* AD 810. Spelt seems to have been the most important cereal at the royal sites in the *Brevium Exempla*, but is generally rare in the archaeobotanical record, unlike free-threshing wheat which exhibits the opposite pattern of archaeobotanical frequency but documentary scarcity (Bakels 2005, 397–398). This discrepancy remains to be fully explained, but it could well relate to status: the documented royal sites may have had preferential access to spelt wheat, whereas the excavated sites in Bakels' survey, whose status is not documented, did not enjoy such access.

Turning to animal husbandry, developments in the seventh to ninth centuries can again be seen to have extended far beyond the study regions. At Catholme (Staffordshire), for example, whose occupation sequence spans at least the early seventh to late ninth centuries, there are large ditch systems comparable to the droveway and paddock complexes discussed in Chapter 2 (Hamerow in Losco-Bradley & Kinsley 2002, 126). At Quarrington (Lincolnshire), the construction around the seventh century of an enclosure complex accompanied a pronounced change in herding, visible in the faunal record as an increase in the proportion of sheep bones at the expense of pigs, and a greater emphasis upon mature animals and therefore also upon secondary products (Taylor 2003, 271–272). Sheep bones also come to dominate assemblages from the estate centre at Bishopstone (east Sussex), and the monastic sites at Hartlepool (Co. Durham) and the Prebendal, Aylesbury (Buckinghamshire) (Poole in Thomas 2010, 142–154; Huntley & Rackham in Daniels 2007, 108–123; Jones in Farley & Jones 2012, 97–104). Perhaps most striking of all, however, is the evidence from the early ninth century phase at Flixborough (Lincolnshire), where *styli* and a lead inscription support Blair's interpretation of the settlement as a probable monastery (Blair 2005, 205–209). An abundance of small loomweights from this phase, together with an increased proportion of sheep bones in the faunal record, suggest a growing interest in woollen textiles reminiscent of that witnessed at Brandon, and similar to trends emerging from the excavations at monastic Lyminge (Kent) (Loveluck 2007, 96–102; Dobney *et al.* 2007, 160–165; Thomas 2013, 137).

Yet the tendency towards sheep and wool is neither exclusive nor universal in the seventh to ninth centuries, even among ecclesiastical sites. Ninth-century Flixborough witnessed the introduction of a new, unprecedentedly large breed of cattle, besides the ovine developments mentioned above. Both the size and the greater maturity of cattle in this phase strongly suggest plough teams (Dobney *et al.* 2007, 155–156). On Lindisfarne, meanwhile, there is an exceptional assemblage of juvenile and neonatal cattle bones at the Green Shiel site, which might represent vellum production for the island's seventh- to ninth-century monastic community (O'Sullivan 2001, 42). Extant manuscripts from this period themselves imply the existence of very extensive herds, capable of supplying large quantities of calf (or lamb or kid) hides to monastic scriptoria: a uniquely high-status need (Hamerow 2002, 151–152).

Within the study regions, as beyond, the sites with the clearest evidence for agrarian innovation between the seventh and ninth centuries are also those with the strongest élite and/or ecclesiastical associations. Excavated remains at Ely, Eynsham, North Elmham Park and Great Wakering can all be plausibly associated with documented ecclesiastical establishments of the seventh to ninth centuries, while artefactual evidence for literacy and Christian metalwork suggest that Brandon was also monastic (Blair 2005, 206). It has been suggested that Yarnton may have been a dependency of the nearby minster at Eynsham; while this view lacks direct supporting evidence, the occurrence at Yarnton of Ipswich Ware pottery far beyond its core distribution zone hints in any case at high-status, long-distance trade links (Hey 2004, 90–91). Similarly, the discovery of Ipswich Ware and continental pottery at Dorney suggest that, if it was indeed the site of a market as proposed by the excavators, then it was a market at which high-status goods were traded (Foreman *et al.* 2002). Furthermore, it is likely that many of the proposed innovations in seventh- to ninth-century agriculture required a scale of investment (in both labour and raw materials) and a degree of planning which might have proved impractical without strong and stable lordship: stockyard complexes and watermills, for instance, presumably required great initial outlay.

On the other hand, we should not underestimate the potential significance of peasant-level cooperation and coordination, and collective investment in and management of communal resources such as ploughs or ovens. Demands for food-rents, and the opportunities for market exchange, could well have stimulated surplus production among the *ceorls*. In a pre-industrial world, trade will almost certainly have developed hand-in-hand with farming, each helping to sustain growth in the other, and the former ultimately dependent on the latter (but not necessarily *vice versa*). But trade, like demography, makes an awkward *primum mobile*, not least because its heyday appears to have post-dated the beginnings of agricultural development. According to current interpretations, the *floruit* of economic intensification can be dated to the early- or mid-eighth century, when the emporia were established as large, planned settlements hosting intensive craft production and goods exchange, and when the monetisation of southern and eastern England was consolidated and extended in the Secondary *sceatta* phase (Naylor 2012, 245–253). According to the evidence discussed in this book, agricultural development had been set in train well before this, in the seventh century.

I would rather see developments in trading and farming as mutually supportive twins, both children of the same lordly parent. Early medieval élites, aspiring to Roman greatness in a changed world, increasingly embedded their power and wealth in the soil, and so facilitated, encouraged and necessitated a transformation in farming which in turn fuelled wider economic change. It was these élites – perhaps above all ecclesiastical élites – who had privileged access to the intellectual, cultural, material and technological influences of both Celtic and Carolingian Europe, whose agrarian innovations so often mirror, and maybe inspired, those of Anglo-Saxon England. It may be fanciful to imagine an eighth-century abbot crossing the channel to Francia in search of masons to build him a grain oven 'in the Roman manner that he loved', as Benedict Biscop famously did for his church at Wearmouth (Tyne and Wear) (Bede, *H.A.B.* 5). Nonetheless, it does seem likely that newly consolidated landlords, the Old English 'loaf-wards', were the prime movers of agrarian development from the seventh century onwards. But if they did see further than other men, it was only by standing on the shoulders of peasants. The loaf-wards must necessarily have acted – perhaps as the initial catalyst – amidst a ready, willing and able peasantry whose part in this story remains to be explored as more evidence comes to light. Subsequently, demands and opportunities arising from a growing population and an intensifying trade network both benefited from and contributed to the transformation of farming. The fruits of this transformation – especially but not exclusively grain and wool – were stored, amassed, transported and redistributed in surplus. Their growing importance can be seen to represent a change in the nature of wealth, directly associated with changes in proprietary land-rights, from the seventh century onwards.

Research into Anglo-Saxon agriculture has, until recently, been almost entirely focused upon the history of field systems, to the neglect of the biological minutiae of crops and livestock. I hope that this book has demonstrated not only the great informative potential of a multi-faceted and ever-growing archaeological and palaeoenvironmental dataset, but also the wider significance of this topic in terms of early medieval culture, economy and environment. England in this period must be seen as a society of farmers, and archaeology is increasingly allowing us to do so as never before. The story of Anglo-Saxon England in the long eighth century, from the cornfield to the king, is a story of farming transformed.

Appendix – Gazetteer of Sites

Addenbrooke's Hospital, Cambridge (Hutchison Site)
County: Cambridgeshire
NGR: TL 4622 5535
Main reference: Evans *et al.* 2008
Zooarchaeology: C. Swaysland in Evans *et al.* 2008
Archaeobotany: K. Roberts in Evans *et al.* 2008
Elevation: 17 m AOD
Geology: Lower Chalk, with second and third terrace gravels nearby
Summary: Curvilinear ditch, pit, wells and two post-built structures are dated to between seventh and ninth centuries, with a likely focus in the eighth and ninth. The dating of the post-built structures and some of the wells is circumstantial, based upon their spatial relationship with the ditch, but other features are dated by the presence of Ipswich Ware, and a clunch block in one of the wells has been radiocarbon-dated to cal. AD 660–790.

Alchester (extramural settlement)
County: Oxfordshire
NGR: SP 5715 2095
Main reference: Booth *et al.* 2001
Zooarchaeology: n/a
Archaeobotany: R. Pelling in Booth *et al.* 2001
Elevation: 65 m AOD
Geology: Clays and gravels in the Upper Thames valley
Summary: A largely Roman-period site, but Anglo-Saxon pottery in layers overlying Romano-British features provides a post-Roman date for a botanical sample. Burnt stone in the Saxon layers also provides a circumstantial case for the survival of a Roman corn-drying oven into the post-Roman period. There is little closely datable Anglo-Saxon evidence, but a stamped sherd, the lack of organic-tempered fabrics, and stratigraphic circumstances suggest a fifth- to sixth-century timeframe.

Bancroft
County: Buckinghamshire
NGR: SP 8273 4033
Main reference: Williams & Zeepvat 1994
Zooarchaeology: J. Holmes & K. Rielly in Williams & Zeepvat 1994
Archaeobotany: S. Nye & M. Jones in Williams & Zeepvat 1994
Elevation: 82 m AOD

Geology: Ridge of boulder clay, and glacial sand and gravel deposits
Summary: One *Grubenhaus* thought to be early- to mid-fifth century in date (dated by a small plain hollow-necked biconical vessel), and a large boundary ditch of uncertain date but aligned with early fifth-century burials.

Barnsley Park
County: Gloucestershire
NGR: SP 083 067
Main reference: Webster *et al.* 1985
Zooarchaeology: B. Noddle in Webster *et al.* 1985
Archaeobotany: n/a
Elevation: 140 m AOD
Geology: Cotswold limestone overlain in places with boulder clay
Summary: Imperfectly understood 'Sub-Roman' phase at the site of a Roman villa.

Barrow Hills (Radley)
County: Oxfordshire
NGR: SU 5135 9815
Main reference: Chambers & McAdam 2007
Zooarchaeology: L. Barnetson in Chambers & McAdam 2007
Archaeobotany: L. Moffett in Chambers & McAdam 2007
Elevation: 60 m AOD
Geology: Second gravel terrace of the Upper Thames
Summary: 45 *Grubenhäuser* and 22 post-built structures, plus inhumations, pits, and fills of prehistoric barrow ditches. These features are generally dated to between the fifth and early seventh centuries by a ceramic assemblage including a variety of fabrics and decorative techniques such as stamping and incising.

Barton Court Farm
County: Oxfordshire
NGR: SU 510 978
Main reference: Miles 1986
Zooarchaeology: B. Wilson in Miles 1986
Archaeobotany: M. Jones & M. Robinson in Miles 1986
Elevation: 60 m AOD
Geology: Second gravel terrace of the Upper Thames
Summary: Anglo-Saxon occupation at a Roman villa site, including Saxon sherds within the fill of a Late Roman ditch,

suggesting that post-Roman occupation had begun before the ditches had silted up. The Anglo-Saxon pottery includes some stamped decoration, supporting an early date, and inhumations at the site include grave goods, such as beads, thought to be of sixth-century date. Overall, a fifth to sixth-century date seems most likely for the post-Roman sequence.

Beech House Hotel, Dorchester-on-Thames
County:	Oxfordshire
NGR:	SU 577 944
Main reference:	Rowley & Brown 1981
Zooarchaeology:	A. Grant in Rowley & Brown 1981
Archaeobotany:	n/a
Elevation:	50 m AOD
Geology:	First gravel terrace of the Upper Thames
Summary:	Post-Roman deposits at an otherwise Roman site, including a faunal assemblage. The deposits are not closely dated but would be consistent with a fifth- to sixth-century chronology.

Benson (St Helen's Avenue)
County:	Oxfordshire
NGR:	SU 61590 91550
Main reference:	Pine & Ford 2003
Zooarchaeology:	S. Hamilton-Dyer in Pine & Ford 2003
Archaeobotany:	Robinson n.d.
Elevation:	49 m AOD
Geology:	First (floodplain) gravel terrace of the Upper Thames
Summary:	Three *Grubenhäuser* and associated postholes, plus two rectilinear ditched enclosures, belong to the Anglo-Saxon phase, which may be dated to the seventh century. The Anglo-Saxon pottery is dominated by organic-tempered wares, and a few sherds bear incised or stamped decoration. Bone from one *Grubenhaus* was radiocarbon-dated to cal. AD 545–659, and also sharply rectilinear enclosures are thought not to occur before the seventh century.

Berinsfield (Mount Farm)
County:	Oxfordshire
NGR:	SU 5822 9661
Main reference:	Lambrick 2010
Zooarchaeology:	B. Wilson in Lambrick 2010
Archaeobotany:	M. Jones & M. Robinson in Lambrick 2010
Elevation:	58 m AOD
Geology:	In a 'saucer of Gault Clay' on the second gravel terrace of the Upper Thames
Summary:	A multi-period site with scattered traces of Anglo-Saxon activity, including a waterhole, pits, and a wattle-lined well. Wooden stakes from the well have been radiocarbon-dated to cal. AD 430–770. Chronological indicators are otherwise limited to annular and intermediate loomweights, and undiagnostic handmade Anglo-Saxon pottery.

Bishop's Cleeve (Church Road)
County:	Gloucestershire
NGR:	SO 95855 27560
Main reference:	Lovell *et al.* 2007
Zooarchaeology:	A. Powell & S. Knight in Lovell *et al.* 2007
Archaeobotany:	R. Pelling in Lovell *et al.* 2007
Elevation:	56 m AOD
Geology:	A deposit of Cheltenham Sand below Cleeve Hill (the highest point in the Cotswolds)
Summary:	Ditches, pits, a posthole and a channel fill, dated by handmade Early/Middle Saxon pottery in various fabrics. A relatively high proportion of organic-tempered wares in the ceramic assemblage may suggest a chronology centred on the seventh century or later.

Bletchley (Water Eaton)
County:	Buckinghamshire
NGR:	SP 8805 3262
Main reference:	Hancock 2010
Zooarchaeology:	J. Rackham in Hancock 2010
Archaeobotany:	G. Martin in Hancock 2010
Elevation:	75 m AOD
Geology:	Gravel terraces of the Ouzel
Summary:	One *Grubenhaus*, pits, ditched enclosures and a trackway are dated to the seventh to ninth centuries by the occurrence, within a small ceramic assemblage, of one sherd of Ipswich Ware and eight of Maxey-type ware.

Bloodmoor Hill (Carlton Colville)
County:	Suffolk
NGR:	TM 5185 8995
Main reference:	Lucy *et al.* 2009
Zooarchaeology:	L. Higbee in Lucy *et al.* 2009
Archaeobotany:	R. Ballantyne in Lucy *et al.* 2009
Elevation:	9 m AOD
Geology:	Sands, in the vicinity of a clay ridge
Summary:	38 *Grubenhäuser*, at least nine post-built structures, four possible middens, around 270 pits, and five hearth/oven bases. Occupation may span the late fifth to early eighth centuries. A developmental sequence is proposed, based upon ceramic and artefactual evidence, and extensive radiocarbon dating: Phase 1 (AD 500–580), Phase 2a (AD 580–650), and Phase 2b (AD 650–700). Unfortunately, neither the faunal nor the botanical materials from this site are clearly assigned to the sub-phases within this sequence.

Brandon (Staunch Meadow)

County:	Suffolk
NGR:	TL 778 864
Main reference:	Tester *et al.* 2014
Zooarchaeology:	P. Crabtree in Tester *et al.* 2014
Archaeobotany:	P. Murphy & V. Fryer in Tester *et al.* 2014
Elevation:	6 m AOD
Geology:	Sand ridge surrounded by peat
Summary:	34 post-built structures, fence lines, pits, ditches, hearths, a church and a cemetery together constitute what is likely to have been a monastic settlement. Dating evidence includes Ipswich Ware, *sceattas* (*c.* AD 720–760) and radiocarbon dates; the main period of occupation activity is dated to between the mid-seventh and mid-ninth centuries. Earlier activity is also suggested by a line of stakes at the south end of the site, radiocarbon-dated to cal. AD 540–650.

Brandon Road North, Thetford

County:	Norfolk
NGR:	TL 855 832
Main reference:	Atkins & Connor 2010
Zooarchaeology:	I. Baxter in Atkins & Connor 2010
Archaeobotany:	V. Fryer in Atkins & Connor 2010
Elevation:	11 m AOD
Geology:	Sands and gravels
Summary:	Seven *Grubenhäuser*, ovens, pits, and a possible hall belong to an Early Saxon phase, centring on the late fifth to early sixth centuries. The dating of this phase is supported by the ceramic assemblage: among the decorated sherds, bossing and incision predominate over stamping. Subsequently, a large enclosure, post-built structure and more ovens were constructed between the seventh and ninth centuries. This phase is dated by the occurrence of Ipswich and Maxey-type wares.

Brettenham (Melford Meadows)

County:	Norfolk
NGR:	TL 878 826
Main reference:	Mudd 2002
Zooarchaeology:	A. Powell & K. M. Clarke in Mudd 2002
Archaeobotany:	M. Robinson in Mudd 2002
Elevation:	13 m AOD
Geology:	Low sandy terrace ridge above the Thet
Summary:	Scatter of eleven *Grubenhäuser*, pits, hollows and hearths. A fifth to sixth century date is suggested by the ceramic specialist, noting the high proportion of sandy wares, and the occurrence of stamping, incised chevrons, a miniature biconical jar with slashed carination, and annular loomweights.

Chadwell St Mary (County Primary School)

County:	Essex
NGR:	TQ 6450 7854
Main reference:	Lavender 1998
Zooarchaeology:	n/a
Archaeobotany:	V. Fryer & P. Murphy in Lavender 1998
Elevation:	30 m AOD
Geology:	Gravel terrace
Summary:	One *Grubenhaus* with a ceramic assemblage dominated by organic-tempered wares, and including a carinated shoulder sherd. A sixth century date has been suggested, but the material would seem to be consistent with a late sixth or early seventh century date, or possibly even later.

Cherry Hinton (Church End, Rosemary Lane)

County:	Cambridgeshire
NGR:	TL 485 576
Main reference:	Mortimer 2003
Zooarchaeology:	C. Swaysland in Mortimer 2003
Archaeobotany:	K. Roberts in Mortimer 2003
Elevation:	15 m AOD
Geology:	Lower Chalk
Summary:	Ditches and shallow pits or hollows, dated by Ipswich Ware to *c.* AD 725–850.

Chieveley

County:	Berkshire
NGR:	SU 479 727
Main reference:	Mudd 2007
Zooarchaeology:	n/a
Archaeobotany:	V. Fryer in Mudd 2007
Elevation:	108 m AOD
Geology:	Mixed sands, clays and gravels of the Reading Beds, in the drainage basin of the Lambourn
Summary:	Groups of pits, somewhat tenuously dated by a small assemblage of organic-tempered pottery to between the sixth and ninth centuries.

Childerley Gate

County:	Cambridgeshire
NGR:	TL 359 598
Main reference:	Abrams & Ingham 2008
Zooarchaeology:	K. Rielly in Abrams & Ingham 2008
Archaeobotany:	J. Giorgi in Abrams & Ingham 2008
Elevation:	65 m AOD
Geology:	Boulder clay and degraded chalk outcrops in the Bourn valley
Summary:	A 'Sub-Roman' phase, in which a 'dark earth' overlies fourth-century features. The ceramic assemblage includes types described as 'very late Roman' and grog-tempered 'proto-Saxon' sherds.

Cogges (Cogges Manor Farm, Witney)

County:	Oxfordshire
NGR:	SP 3621 0963

Main reference:	Rowley & Steiner 1996
Zooarchaeology:	n/a
Archaeobotany:	M. Robinson in Rowley & Steiner 1996
Elevation:	81 m AOD
Geology:	An outcrop of Cornbrash formation limestone above the Windrush floodplain
Summary:	Pits, postholes and a *Grubenhaus* may be dated to the fifth to sixth centuries, by the occurrence of stamped and incised decoration on sherds, most of which are of calcareous fabrics. Botanical samples ('presence' data only) derive from the fills of the postholes and the *Grubenhaus*.

Collingbourne Ducis (Cadley Road)

County:	Wiltshire
NGR:	SU 24450 54000
Main reference:	Pine 2001
Zooarchaeology:	S. Hamilton-Dyer in Pine 2001
Archaeobotany:	J. Letts in Pine 2001
Elevation:	132 m AOD
Geology:	Upper Chalk, with river gravels to the west of the site
Summary:	Ten *Grubenhäuser*, plus pits and postholes, producing charred plant material and some faunal remains. The ceramic assemblage is dominated by organic-tempered sherds. Radiocarbon dates from the bone from the *Grubenhaus* fills suggest that the main occupation sequence fell within the seventh to ninth centuries, although material from one *Grubenhaus* returned an earlier determination of cal. AD 430–660.

Cottenham (Lordship Lane)

County:	Cambridgeshire
NGR:	TL 449 681
Main reference:	Mortimer 1998; 2000
Zooarchaeology:	L. Higbee in Mortimer 1998
Archaeobotany:	C. Stevens in Mortimer 1998
Elevation:	10 m AOD
Geology:	Ridge of Lower Greensand, overlying Kimmeridge Clay
Summary:	An extensive ditch system with post-built structures and one *Grubenhaus* constitute an initial phase. Subsequently, new enclosures were created, apparently representing radial growth from an unseen core to the south-east of the excavated site. Varying proportions of Ipswich Ware in the respective ceramic assemblages suggest that the first phase is of seventh to eighth century date, with the second phase falling within the eighth to ninth centuries.

Cresswell Field

County:	Oxfordshire
NGR:	SP 470 114

Main reference:	Hey 2004
Zooarchaeology:	J. Mulville & K. Ayres in Hey 2004
Archaeobotany:	R. Pelling in Hey 2004
Elevation:	62 m AOD
Geology:	Second gravel terrace of the Upper Thames
Summary:	Rectangular post-built structure, fenced enclosures, and at least three *Grubenhäuser*. A bone comb from the post-built structure has been radiocarbon-dated to cal. AD 640–810, and cereal grains and charcoal from other postholes have been radiocarbon-dated to cal. AD 770–1010, 660–890, and 660–900 respectively. The structure is thus believed to be of eighth to ninth century date. One of the *Grubenhäuser* appears to represent earlier activity: a cattle skull from one returned radiocarbon dates of cal. AD 410–650 and 540–680. Another, however, may have been contemporary with the post-built structure since bone in its backfill return radiocarbon determinations of cal. AD 640–860 and 530–680.

Criminology Site, Cambridge (Sidgwick Campus)

County:	Cambridgeshire
NGR:	TL 4428 5812
Main reference:	Dodwell *et al.* 2004
Zooarchaeology:	C. Swaysland in Dodwell *et al.* 2004
Archaeobotany:	K. Roberts in Dodwell *et al.* 2004
Elevation:	9 m AOD
Geology:	Second gravel terrace of the Cam
Summary:	A post-built structure, pits and two *Grubenhäuser*. The ceramic assemblage includes one stamped and one incised sherd, and organic-tempered fabrics constitute 40% of the assemblage by weight. This assemblage is considered to be consistent with a sixth to seventh century date.

Didcot (Milton Park)

County:	Oxfordshire
NGR:	SU 4970 9235
Main reference:	Williams 2008
Zooarchaeology:	n/a
Archaeobotany:	L. Evans in Williams 2008
Elevation:	57 m AOD
Geology:	Second gravel terrace of Upper Thames
Summary:	Four botanical samples (without fully quantified data), potentially associated with Early Saxon activity. The associated ceramic assemblage, which includes seven stamped sherds, is taken to indicate a date between the fifth and seventh centuries, with activity perhaps concentrated in the sixth century. The dating of the samples is here considered tenuous.

Downham Market Bypass

County:	Norfolk
NGR:	TF 6097 0211
Main reference:	Percival 2001
Zooarchaeology:	n/a
Archaeobotany:	n/a
Elevation:	22 m AOD
Geology:	Sands and clays, near the margin between the silt and peat fens
Summary:	Sequence of pits and ditches, dated by Ipswich Ware to *c.* AD 725–850.

Downham Market: London Road

County:	Norfolk
NGR:	TF 610 022
Main reference:	Trimble 2001
Zooarchaeology:	n/a
Archaeobotany:	n/a
Elevation:	21 m AOD
Geology:	Sands and clays, near the margin between the silt and peat fens
Summary:	Sequence of ditches, dated by Ipswich Ware to *c.* AD 725–850.

Dumbleton (Bank Farm)

County:	Gloucestershire
NGR:	SP 0272 3658
Main reference:	Coleman *et al.* 2006
Zooarchaeology:	n/a
Archaeobotany:	W.J. Carruthers in Coleman *et al.* 2006
Elevation:	60 m AOD
Geology:	Thin gravel layer overlying clay – probably a remnant of the second terrace of the Avon
Summary:	Ditch fills yielding generic Early/Mid Saxon pottery (fifth to ninth centuries), and one associated botanical sample. However, charred plant remains in other samples – considered to be of prehistoric date – returned radiocarbon dates spanning the sixth to tenth centuries.

Duxford

County:	Cambridgeshire
NGR:	TL 4805 4585
Main reference:	Lyons 2011
Zooarchaeology:	I. Baxter in Lyons 2011
Archaeobotany:	n/a
Elevation:	29 m AOD
Geology:	Chalk
Summary:	Three *Grubenhäuser*, a post-built structure, pits, a gully and fence-line, whose positioning respects a Romano-British 'drying building' which may therefore have continued to stand into the early medieval period. Dating is uncertain: annular loomweights and various handmade pottery fabrics would be consistent with a fifth- to seventh-century date, but some shelly sherds

resemble seventh- to ninth-century Maxey-type ware.

Eaton Socon (Alpha Park, Bell Farm, Great North Road)

County:	Cambridgeshire
NGR:	TL 168 581
Main reference:	Hood 2007
Zooarchaeology:	n/a
Archaeobotany:	W. Smith in Hood 2007
Elevation:	19 m AOD
Geology:	River terrace gravels and sands
Summary:	One *Grubenhaus* whose fill contained 20 sherds of handmade 'Early Saxon' pottery of sandy fabric. One rim sherd of a globular vessel with marked shoulder, wide mouth and upright rim is considered analogous to a form dated by Myres to the fifth century. Two supposedly curated Romano-British sherds derive from the same fill.

Ely: Ashwell Site (West Fen Road)

County:	Cambridgeshire
NGR:	TL 529 808
Main reference:	Mortimer *et al.* 2005
Zooarchaeology:	L. Higbee in Mortimer *et al.* 2005
Archaeobotany:	R. Ballantyne in Mortimer *et al.* 2005
Elevation:	9 m AOD
Geology:	Kimmeridge Clay; some sand and sandstone elsewhere on the island of Ely
Summary:	Ditched enclosures, pits, post-built structures and a possible trackway. The dominance of Ipswich Ware in the ceramic assemblage, along with a *sceatta* dated *c.* AD 730–740, suggests an eighth- to ninth-century chronology.

Ely: Chiefs Street

County:	Cambridgeshire
NGR:	TL 5356 8042
Main reference:	Kenney 2002
Zooarchaeology:	I. Baxter in Kenney 2002
Archaeobotany:	C. Stevens in Kenney 2002
Elevation:	21 m AOD
Geology:	Lower Greensand at the highest point of the island of Ely
Summary:	Pits, wells and an oven dated to the seventh to ninth centuries by a ceramic assemblage consisting mostly of Ipswich Ware, but also including some Maxey-type ware.

Ely: Consortium Site (West Fen Road)

County:	Cambridgeshire
NGR:	TL 531 809
Main reference:	Mudd & Webster 2011
Zooarchaeology:	L. Higbee in Mudd & Webster 2011
Archaeobotany:	W. J. Carruthers in Mudd & Webster 2011
Elevation:	6 m AOD

Geology: Kimmeridge Clay; some sand and sandstone elsewhere on the island of Ely.

Summary: A series of shallow ditched enclosures, plus pits, a possible well, and gullies and posthole alignments that hint at structures. Maxey-type and Ipswich Ware dominate the ceramic assemblage, suggesting that occupation centred on the eighth to ninth centuries. A *sceatta* dated *c.* AD 740–750, and loomweights resembling the 'intermediate' type, lend further support to this chronology.

Eye (Hartismere High School)

County: Suffolk
NGR: TM 13800 74040
Main reference: Caruth 2008
Zooarchaeology: Higbee 2009
Archaeobotany: Fryer 2008
Elevation: 39 m AOD
Geology: Heavy boulder clay, with gravel terraces along river valleys in the vicinity
Summary: 19 *Grubenhäuser*, two post-built structures, ditches, pits, layers and a cobbled track with wheel ruts. The settlement is described as being of Early Saxon date, but dating evidence was not available for reference at the time of writing.

Eynesbury

County: Cambridgeshire
NGR: TL 180 585
Main reference: Ellis 2004
Zooarchaeology: N. Sykes in Ellis 2004
Archaeobotany: A. Clapham in Ellis 2004
Elevation: 15 m AOD
Geology: First and second gravel terraces of the Great Ouse
Summary: Seven *Grubenhäuser*, plus pits, postholes and ditches. Annular loomweights, plus pottery bearing bossed, incised, and impressed decoration (and with a relatively low occurrence of organic-tempered sherds), might support a fifth- to sixth-century date.

Eynsham (Abbey)

County: Oxfordshire
NGR: SP 433 091
Main reference: Hardy *et al.* 2003
Zooarchaeology: J. Mulville in Hardy *et al.* 2003
Archaeobotany: R. Pelling in Hardy *et al.* 2003
Elevation: 66 m AOD
Geology: Calcareous gravel terrace, overlying Oxford Clay, at the confluence of the Evenlode and the Thames.
Summary: Early Saxon occupation (Phase 2a) is represented by five or six *Grubenhäuser*, plus a pit, fence line, and postholes. The occurrence of incised and especially stamped decoration on sherds is cited in support of a predominantly sixth-century date. Mid Saxon occupation (2b) is initially represented by pits, burnt areas, hearths and associated postholes and beamslots. Apart from its stratigraphic relationship to Phases 2a and 2c, this phase is dated by two early eighth-century *sceattas* found in a large pit. Later Mid Saxon occupation (2c) is represented two probable post-hole buildings, an alignment of pits, and boundary ditches. Small amounts of Late Saxon pottery and an early to mid-ninth-century Northumbrian *styca* suggest a terminal date in the latter part of the ninth century.

Although the stratigraphic relationships between Phases 2b and 2c features are not all clearly understood, the excavators suggest that the phases may span the mid-seventh to the early or mid-eighth centuries, and the early or mid-eighth to the late ninth centuries respectively.

Flixton (Flixton Park Quarry)

County: Suffolk
NGR: TM 30370 86670
Main reference: Boulter 2008
Zooarchaeology: J. Curl in Boulter 2008
Archaeobotany: V. Fryer in Boulter 2008
Elevation: 17 m AOD
Geology: River terrace gravels
Summary: *Grubenhäuser*, post-built structures, pits, ditches and a putative shrine. Pottery decoration includes fifth- and seventh-century elements but is said to fit best with a sixth-century date: incised and combed lines, stamping, stabbing, corrugation and finger-impression are cited in support of this interpretation.

Forbury House (Reading)

County: Berkshire
NGR: SU 7180 7350
Main reference: Edwards 2008
Zooarchaeology: n/a
Archaeobotany: Vaughan-Williams 2005
Elevation: 42 m AOD
Geology: Taplow Gravel terrace in the Kennet valley, near a confluence with the Thames
Summary: Two pits, one containing a sherd of handmade Anglo-Saxon pottery with a burnished interior; the other producing charred plant remains, plus charcoal radiocarbon-dated to cal. AD 630–960.

Fordham (Hillside Meadow)

County: Cambridgeshire
NGR: TL 632 706

Main reference:	Patrick & Rátkai 2011
Zooarchaeology:	I. Baxter in Patrick & Rátkai 2011
Archaeobotany:	W. Smith in Patrick & Rátkai 2011
Elevation:	15 m AOD
Geology:	River terrace deposits and chalk
Summary:	Three or four *Grubenhäuser* and ditches marking enclosures and a droveway. The ceramic assemblage includes both Early Saxon (stamped sherd) and Mid Saxon (Ipswich Ware) indicators; the morphology of the site would be most consistent with a seventh-century or later date.

Frocester Court

County:	Gloucestershire
NGR:	SO 785 032
Main reference:	Price 2000
Zooarchaeology:	B. Noddle in Price 2000
Archaeobotany:	n/a
Elevation:	40 m AOD
Geology:	Third gravel terrace of the Severn, overlying clay
Summary:	Imperfectly understood 'Sub-Roman' or post-Roman phase at a villa site. A phase which included an ox skull radiocarbon-dated to cal. AD 534–632 was followed by occupation represented by organic-tempered pottery.

Gamlingay (Station Road)

County:	Cambridgeshire
NGR:	TL 2430 5190
Main reference:	Murray 2005
Zooarchaeology:	T. Roberts in Murray 2005
Archaeobotany:	V. Fryer in Murray 2005
Elevation:	47 m AOD
Geology:	Low rise in a Lower Greensand ridge
Summary:	Sequence of enclosed 'farmsteads' comprising ditch systems (enclosures and droveways), *Grubenhäuser*, post-built structures and pits. Annular loomweights and Charnwood Forest pottery may indicate Early Saxon activity, but the site morphology and occurrence of Maxey-type wares would be more consistent with a seventh-century or later date for the enclosed settlement sequence, including features dated to the eighth and ninth centuries by the presence of Ipswich Ware.

Godmanchester (Cardinal Distribution Park)

County:	Cambridgeshire
NGR:	TL 2550 7030
Main reference:	Gibson 2003
Zooarchaeology:	I. Baxter in Gibson 2003
Archaeobotany:	V. Fryer in Gibson 2003
Elevation:	13 m AOD
Geology:	Interface of first and second gravel terraces and Oxford Clay

Summary:	Six *Grubenhäuser*, ditched enclosure system, and traces of post-built structures. The excavators suggest two alternative sequences of three phases of development. The published report favours an Early Saxon date, but the site morphology, the presence of both annular and intermediate loomweights, and a small amount of Ipswich Ware (six sherds from the same segment of a ditch) are arguably more consistent with a seventh- to eighth-century date.

Goring (Gatehampton Farm)

County:	Oxfordshire
NGR:	SU 606 797
Main reference:	Allen *et al.* 1995
Zooarchaeology:	B. Wilson in Allen *et al.* 1995
Archaeobotany:	J. Letts in Allen *et al.* 1995
Elevation:	47 m AOD
Geology:	Thames gravel terrace
Summary:	A midden deposit and *Grubenhaus* are dated to the Anglo-Saxon period by a ceramic assemblage comprising several handmade fabrics; they also produced a tiny faunal assemblage and a charred botanical sample. Additionally, a decorated sherd dated to the late fifth or sixth century derives from a Roman grain oven, which may therefore have survived into the early medieval period.

Great Linford (Church)

County:	Buckinghamshire
NGR:	SP 856 419
Main reference:	Mynard & Zeepvat 1992
Zooarchaeology:	J.M Holmes in Mynard & Zeepvat 1992
Archaeobotany:	P. Busby in Mynard & Zeepvat 1992
Elevation:	90 m AOD
Geology:	Gravel terrace of the Ouse, Oxford Clay beds, and limestone outcrops in the vicinity of the site.
Summary:	Undisturbed buried soil underlying a medieval church, yielding Early and Mid Saxon sherds (including a sherd with incised decoration, and Maxey-type ware). The volume and condition of the sherds suggests that the context represents activity spanning the fifth to ninth centuries. Small faunal and botanical assemblages derive from this context.

Great Wakering (St Nicholas' church)

County:	Essex
NGR:	TQ 9503 8755
Main reference:	Dale *et al.* 2010
Zooarchaeology:	n/a
Archaeobotany:	V. Fryer in Dale *et al.* 2010
Elevation:	5 m AOD
Geology:	Brickearth overlying Barling Gravels terrace

Summary: Possible site of a Mid to Late Saxon minster church. Excavated features include enclosure ditches, a piece of carved stone, and a 'composite hearth' (possibly a grain oven). Shelly fabrics and predominance of organic-tempered ware, along with the enclosure ditches and putative monastic associations, suggest a seventh-century or later date. The carved stone is assigned a late eighth- to tenth-century date, partly on stylistic grounds.

Handford Road, Ipswich
County: Suffolk
NGR: TM 1530 4455
Main reference: Boulter 2005
Zooarchaeology: J. Curl in Boulter 2005
Archaeobotany: V. Fryer in Boulter 2005
Elevation: 5 m AOD
Geology: Sands and gravels
Summary: Five *Grubenhäuser*, at least three post-built structures, postholes, pits and an oven. A fifth- to sixth-century chronology is preferred in the excavation report, based upon the lack of organic-tempered and Ipswich Ware pottery, and the occurrence of several stamped, incised and bossed sherds.

Harston Mill
County: Cambridgeshire
NGR: TL 418 507
Main reference: O'Brien 2016
Zooarchaeology: R. Jones & C. Phillips in O'Brien 2016
Archaeobotany: R. Scaife in O'Brien 2016
Elevation: 15 m AOD
Geology: Chalk and gravelly clay
Summary: An Early to Mid Saxon sequence at a multi-period site, including six *Grubenhäuser*, several pits, and a series of parallel, linear ditches. Some fifth- to seventh-century features, mainly *Grubenhäuser*, are dated by the presence of stamped and grooved handmade pottery and a bone radiocarbon-dated to cal. AD 570–660. Some eighth- to ninth-century features are dated by the occurrence of Ipswich Ware, but many other features could be assigned only a generic fifth- to ninth-century date.

Heybridge (Elms Farm)
County: Essex
NGR: TL 847 082
Main reference: Atkinson & Preston 1998
Zooarchaeology: Johnstone & Albarella 2002
Archaeobotany: n/a
Elevation: 4m AOD
Geology: River terrace gravels

Summary: Postholes, post-built structures and *Grubenhäuser* in a 'Sub-Roman' phase of a Romano-British settlement, for which precise dating evidence could not be obtained.

High Post, Salisbury
County: Wiltshire
NGR: SU 14400 37130
Main reference: Powell 2011
Zooarchaeology: n/a
Archaeobotany: n/a
Elevation: 131 m AOD
Geology: Upper Chalk with bands of clay-with-flints
Summary: Romano-British masonry oven with T-shaped flue. Post-Roman usage could potentially be indicated by the presence of a pedestal base of fifth- to sixth-century type, and charred material from the base of the flue radiocarbon-dated to cal. AD 335–535.

Hintlesham (Silver Birches, Silver Hill)
County: Suffolk
NGR: TM 0925 4340
Main reference: Boulter 2010
Zooarchaeology: n/a
Archaeobotany: V. Fryer in Boulter 2010
Elevation: 46 m AOD
Geology: Sand and clay
Summary: A ditch dated to c.AD 725–850 by the presence of Ipswich Ware.

Hinxton Hall
County: Cambridgeshire
NGR: TL 496 448
Main reference: Gaimster *et al.* 1998 p.119
Zooarchaeology: L. Gidney, cited by Baxter in Patrick & Rátkai 2011 p.86
Archaeobotany: n/a
Elevation: 37 m AOD
Geology: Chalk and terrace gravels of the Cam
Summary: Four *Grubenhäuser* and a timber hall, producing a faunal assemblage. Detailed dating evidence could not be obtained, but the site is described as 'Early Saxon'.

Hinxton Quarry
County: Cambridgeshire
NGR: TL 487 466
Main reference: Mortimer & Evans 1996
Zooarchaeology: E. Yannouli in Mortimer & Evans 1996
Archaeobotany: C. Stevens in Mortimer & Evans 1996
Elevation: 25 m AOD
Geology: Gravels
Summary: Pits and two *Grubenhäuser*, producing a small faunal assemblage and a single botanical sample. The ceramic assemblage includes a single stamped sherd, plus

short-necked vessels with sixth-century parallels.

Hungerford (Charnham Lane)

County:	Berkshire
NGR:	SU 335 692
Main reference:	Ford 2002
Zooarchaeology:	J. Lovett in Ford 2002
Archaeobotany:	W.J. Carruthers in Ford 2002
Elevation:	100 m AOD
Geology:	Peat, clay, silt and gravel on the floodplain and terrace of the Kennet
Summary:	Faunal remains from a pit and very sparse plant remains from a *Grubenhaus*. Sandy and organic-tempered potsherds point to a sixth- to ninth-century date.

Ingleborough (West Walton)

County:	Norfolk
NGR:	TF 4727 1481
Main reference:	Crowson *et al.* 2005
Zooarchaeology:	P. Baker in Crowson *et al.* 2005
Archaeobotany:	P. Murphy in Crowson *et al.* 2005
Elevation:	3 m AOD
Geology:	Silt roddon (a raised fenland island)
Summary:	Pits, ditches, gullies and a hearth, dated by Ipswich Ware – plus a sherd of North French Blackware and one of possible Tating Ware – to the eighth to ninth centuries.

Ipswich

County:	Suffolk
NGR:	TM 162 441
Main reference:	Scull 2009 pp.313–319 summarises current interpretations
Zooarchaeology:	P. Crabtree 2012
Archaeobotany:	P. Murphy 2004
Elevation:	15 m AOD
Geology:	Clay, sands and gravels
Summary:	Series of rescue excavations spanning 1974–1988, representing the Mid Saxon 'wic' or emporium of *Gipeswic*. One botanical sample derives from a ditch of late sixth- or seventh-century date. Other material is dated by a range of evidence to the seventh to ninth centuries, when the wic flourished.

Kilverstone

County:	Norfolk
NGR:	TL 8840 8385
Main reference:	Garrow *et al.* 2006
Zooarchaeology:	L. Higbee in Garrow *et al.* 2006
Archaeobotany:	R. Ballantyne in Garrow *et al.* 2006
Elevation:	17 m AOD
Geology:	Sand with flint nodules and patches of chalk
Summary:	At least ten *Grubenhäuser*, four post-built structures, and pits occupying the

site of a Romano-British settlement. 'Early Saxon' pottery, including stamped and incised decoration, would be consistent with activity centred on the sixth century. The environmental reports warn that residual Roman material may be present in the Anglo-Saxon contexts.

Lackford Bridge Quarry

County:	Suffolk
NGR:	TL 791 713
Main reference:	Tipper 2007
Zooarchaeology:	n/a
Archaeobotany:	P. Murphy in West 1985
Elevation:	16 m AOD
Geology:	Sands and gravels
Summary:	A post-built structure, two *Grubenhäuser*, pits and postholes. Botanical samples derive from pits and a *Grubenhaus* which is not closely datable within the fifth- to ninth-century period. One of the pits, however, contained charcoal radiocarbon-dated to cal. AD 640–900.

Lake End Road, Dorney

County:	Buckinghamshire
NGR:	SU 9294 7960
Main reference:	Foreman *et al.* 2002
Zooarchaeology:	A. Powell & K. Clark in Foreman *et al.* 2002
Archaeobotany:	R. Pelling in Foreman *et al.* 2002
Elevation:	22 m AOD
Geology:	First terrace gravels, within a system of silted up palaeochannels
Summary:	Pits, whose fills are dated by the presence of Ipswich Ware, North French Ware, Tating Ware, and radiocarbon determinations to the seventh to ninth centuries – perhaps centring specifically on the mid-eighth century. However, charred grain from one pit was radiocarbon-dated to cal. AD 430–660, despite the occurrence of eighth-century pottery in the same context. This site comprises two adjacent excavations, Lake End Road East and West, both of which are close to Lot's Hole (discussed below).

Lakenheath: RAF Lakenheath Consolidated Support and Family Complexes

County:	Suffolk
NGR:	TL 7341 8046
Main reference:	Caruth 2006
Zooarchaeology:	J. Curl in Caruth 2006
Archaeobotany:	V. Fryer in Caruth 2006
Elevation:	10 m AOD
Geology:	Fine Breckland sand
Summary:	Ditches, pits, postholes and a soil layer dated by Ipswich Ware to *c.* AD 725–850.

Lakenheath: RAF Lakenheath Dental Clinic
County:	Suffolk
NGR:	TL 7327 8050
Main reference:	Tester 2006
Zooarchaeology:	A. Willett in Tester 2006
Archaeobotany:	n/a
Elevation:	10 m AOD
Geology:	Fine Breckland sand
Summary:	Three *Grubenhäuser*, ditch systems, wells, pits and postholes producing potsherds of 'sixth- to seventh-century' date. Material from this phase also produced a radiocarbon date of cal. AD 430–640. Further pits and ditches were dated by the presence of Ipswich Ware to *c.* AD 725–850.

Latimer
County:	Buckinghamshire
NGR:	SU 998 986
Main reference:	Branigan 1971
Zooarchaeology:	R. Hamilton in Branigan 1971
Archaeobotany:	n/a
Elevation:	95 m AOD
Geology:	Clay-with-flints, glacial gravels, and alluvium in the Chess Valley
Summary:	Imperfectly understood 'Sub-Roman' phase at a Roman villa site.

Latton Quarry
County:	Wiltshire
NGR:	SU 083 956
Main reference:	Pine 2009
Zooarchaeology:	M. Holmes in Pine 2009
Archaeobotany:	L. Cramp in Pine 2009
Elevation:	82 m AOD
Geology:	Gravel terrace of the Upper Thames
Summary:	Pits and a post-built structure, whose foundations include both postholes and trenches – the latter are more common from the seventh century onwards. Bone from a posthole was radiocarbon-dated to cal. AD 406–544, while a placed deposit of a cow skull in a pit was radiocarbon-dated to cal. AD 501–618.

Lechlade (Sherborne House)
County:	Gloucestershire
NGR:	SU 2126 9974
Main reference:	Bateman *et al.* 2003
Zooarchaeology:	M. Maltby in Bateman *et al.* 2003
Archaeobotany:	C. Stevens in Bateman *et al.* 2003
Elevation:	77 m AOD
Geology:	Second gravel terrace of the Upper Thames
Summary:	Six *Grubenhäuser*, three post-built structures, ditched enclosures and a possible trackway. On morphological grounds, the sequence of successive ditched features is likely to date from the seventh century or later. The high proportion of organic-tempered sherds in the ceramic assemblage would also be consistent with such a chronology. The faunal assemblage derives almost exclusively from the *Grubenhäuser*, and four botanical samples all derive from the same *Grubenhaus*.

Littlemore (Oxford Science Park)
County:	Oxfordshire
NGR:	SP 5390 0210
Main reference:	Moore 2001
Zooarchaeology:	C. Ingrem in Moore 2001
Archaeobotany:	R. Pelling in Moore 2001
Elevation:	63 m AOD
Geology:	Sands and sandy clays of the Corallian limestone ridge, with alluvium along the nearby Littlemore Brook
Summary:	A string of at least ten *Grubenhäuser* covering 265 m. These are dated to the sixth and possibly seventh centuries by the ceramic assemblage recovered from their fills, including handmade 'Early/Mid Saxon' pottery with some organic-tempered sherds, and some bearing stamped or incised decoration. Some pits may also be of Anglo-Saxon date, but their fills are less securely dated. The vast majority of Anglo-Saxon animal bones and plant remains derive from the *Grubenhäuser*.

Lot's Hole, Dorney
County:	Buckinghamshire
NGR:	SU 9220 7970
Main reference:	Foreman *et al.* 2002
Zooarchaeology:	A. Powell & K. Clark in Foreman *et al.* 2002
Archaeobotany:	R. Pelling in Foreman *et al.* 2002
Elevation:	23 m AOD
Geology:	First terrace gravels, within a system of silted up palaeochannels
Summary:	Pits, whose fills are dated by the presence of Ipswich Ware, North French Ware, Tating Ware, and radiocarbon determinations to the seventh to ninth centuries – perhaps centring specifically on the mid-eighth century. This site is close to Lake End Road, discussed above.

Lower Cambourne
County:	Cambridgeshire
NGR:	TL 31080 59460
Main reference:	Wright *et al.* 2009
Zooarchaeology:	n/a
Archaeobotany:	C. Stevens in Wright *et al.* 2009
Elevation:	63 m AOD
Geology:	Boulder clay

Summary: Poorly represented post-Roman activity consisting mainly of pits, wells, and possibly the upper fills of a Roman ditch. The contexts are not closely datable within the fifth- to ninth-century period. Two botanical samples derive from a pit and ditch fill respectively, but it is possible that the macrofossils may represent residual earlier material.

Lower Icknield Way
County: Buckinghamshire
NGR: SP 893 126
Main reference: Masefield 2008
Zooarchaeology: L. Sibun in Masefield 2008
Archaeobotany: R. Scaife in Masefield 2008
Elevation: 108 m AOD
Geology: Gault Clay of the Vale of Aylesbury, with chalk formations immediately to the south
Summary: Pits, and possibly an enclosure ditch, dated to the seventh century (or possibly later) by a ceramic assemblage with a high proportion of organic-tempered sherds.

Lower Slaughter (Copsehill Road)
County: Gloucestershire
NGR: SP 1650 2267
Main reference: Kenyon & Watts 2006
Zooarchaeology: E. Hambleton in Kenyon & Watts 2006
Archaeobotany: J. Jones in Kenyon & Watts 2006
Elevation: 138 m AOD
Geology: Gravel terrace to the north of the Slaughter Brook
Summary: Sequence of shallow ditches containing Late Roman and Anglo-Saxon pottery, plus bone with radiocarbon dates converging on cal. AD 660–900.

Marham (Old Bell)
County: Norfolk
NGR: TF 7084 0979
Main reference: Newton 2010
Zooarchaeology: J. Morris & S. Leach in Newton 2010
Archaeobotany: R. Scaife in Newton 2010
Elevation: 11 m AOD
Geology: Spring line between Lower and Middle Chalk
Summary: Backfill of a *Grubenhaus*, which may be tentatively dated to the ninth century by the presence of Ipswich and Thetford-type ware, a bone comb of ninth-century (or later) type, and a bun-shaped loomweight.

Market Lavington (Grove Farm)
County: Wiltshire
NGR: SU 0135 5415
Main reference: Williams & Newman 2006
Zooarchaeology: J. Bourdillon in Williams & Newman 2006

Archaeobotany: V. Straker in Williams & Newman 2006
Elevation: 100 m AOD
Geology: Greensand ridge between the chalk plateau of Salisbury Plain to the south and Gault Clays to the north
Summary: Ditches, gullies, stakeholes, pits, three *Grubenhäuser*, and a possible beamslot. The latter feature and the predominance of organic-tempered pottery suggest occupation centred on the seventh century or possibly later.

Mucking
County: Essex
NGR: TQ 673 803
Main reference: Hamerow 1993
Zooarchaeology: G. Done in Hamerow 1993
Archaeobotany: M. van der Veen in Hamerow 1993
Elevation: 30 m AOD
Geology: Gravel terrace in the Lower Thames
Summary: 203 *Grubenhäuser*, 53 post-built structures, pits, ditches and hearths in a 'drifting' settlement sequence. Detailed examination of the ceramic assemblage, in conjunction with some metalwork in the fills of the *Grubenhäuser*, suggested three main phases of occupation dating to the fifth century, the fifth to sixth centuries, and the sixth to early eighth centuries respectively.

Neptune Wood
County: Oxfordshire
NGR: SU 550 937
Main reference: Allen *et al.* 2010
Zooarchaeology: F. Worley in Allen *et al.* 2010
Archaeobotany: R. Pelling in Allen *et al.* 2010
Elevation: 51 m AOD
Geology: First gravel terrace of the Upper Thames
Summary: A group of pits containing undiagnostic Anglo-Saxon pottery. A faunal assemblage derives from two of the pits, and two botanical samples derive from separate layers within a large pit. Charred barley grains from the samples have been radiocarbon-dated to cal. AD 590–670.

New Wintles Farm (near Eynsham)
County: Oxfordshire
NGR: SP 432 108
Main reference: Gray 1974
Zooarchaeology: Noddle 1975
Archaeobotany: n/a
Elevation: 64 m AOD
Geology: Gravel terrace in the Upper Thames valley
Summary: *Grubenhäuser*, post-built structures, pits and fence lines, representing at least two stages of settlement at this largely unpublished site. A group of features to the south is dated to the sixth century, and another to the north is dated to

between the seventh and (at least) early eighth centuries. Dating is based on an eighth-century metal object, and Berisford's interpretation of the ceramic assemblage.

North Elmham Park

County:	Norfolk
NGR:	TF 987 215
Main reference:	Wade-Martins 1980
Zooarchaeology:	B. Noddle in Wade-Martins 1980
Archaeobotany:	H. Jarman in Wade-Martins 1980
Elevation:	45 m AOD
Geology:	Plateau gravel over the chalky boulder clay
Summary:	A large settlement site with probable ecclesiastical associations, including ditch systems, buildings with substantial foundation trenches, wells and pits. Eighth- to ninth-century dates are provided by Ipswich Ware, one sherd of Tating Ware, and dendrochronology of timbers from the well (AD 793–795).

Old Castle Inn, Dorchester-on-Thames

County:	Oxfordshire
NGR:	SU 578 942
Main reference:	Bradley 1978
Zooarchaeology:	A. Grant in Bradley 1978
Archaeobotany:	n/a
Elevation:	50 m AOD
Geology:	A faunal assemblage from post-Roman deposits of imprecise date, following Roman-period occupation at this site.

Old Windsor

County:	Berkshire
NGR:	SU 991 746
Main reference:	Wilson & Hurst 1958
Zooarchaeology:	n/a
Archaeobotany:	n/a
Elevation:	17 m AOD
Geology:	Floodplain gravel terrace of the Middle Thames
Summary:	Remains of a triple vertical watermill and leat. Dendrochronology of surviving timbers suggests a probable construction date in the late seventh century.

Orton Hall Farm

County:	Cambridgeshire
NGR:	TL 1765 9555
Main reference:	Mackreth 1996
Zooarchaeology:	J. King in Mackreth 1996
Archaeobotany:	n/a
Elevation:	20 m AOD
Geology:	Oxford Clay and river terrace gravels
Summary:	Complicated transitional 'Anglo-Saxon' phase at a Roman villa site, extending into the fifth or sixth century, dated by a comb of late fourth- or early fifth-century

type, late fifth- or early sixth-century copper alloy plates, and pottery bearing stamped, bossed and grooved decoration.

Orton Waterville (Cherry Orton Road)

County:	Cambridgeshire
NGR:	TL 1568 9627
Main reference:	Wright 2004
Zooarchaeology:	S. Knight in Wright 2004
Archaeobotany:	C. Stevens in Wright 2004
Elevation:	16 m AOD
Geology:	Gravel terrace of the Nene
Summary:	Two pits (one of which may represent a *Grubenhaus*) and an arrangement of ditches. The ceramic assemblage from these features lacks diagnostic forms, fabrics or decoration but – together with the linear ditch systems – would be consistent with a seventh-century or later date.

Outwell (Church Terrace)

County:	Norfolk
NGR:	TF 514 037
Main reference:	Hall 2003
Zooarchaeology:	n/a
Archaeobotany:	V. Fryer in Hall 2003
Elevation:	4 m AOD
Geology:	Clays, near the margin between the silt and peat fens
Summary:	A cut which may represent an earlier, wider course of the Nene. Material from this feature has been dated to the late fifth or early sixth century by a ceramic assemblage including quartz- and organic-tempered pottery with incised patterns.

Pampisford (Bourn Bridge)

County:	Cambridgeshire
NGR:	TL 516 495
Main reference:	Pollard 1996
Zooarchaeology:	E. Yannouli in Pollard 1996
Archaeobotany:	V. Fryer & P. Murphy in Pollard 1996
Elevation:	27m AOD
Geology:	Gravel terrace of the Granta, a tributary of the Cam
Summary:	Anglo-Saxon settlement within relict Romano-British field system, consisting of *Grubenhäuser*, pits, hollows and – on the floodplain to the north-west – a post-built structure which is said to be of seventh-century type. Distinctively Early or Mid Saxon pottery types are lacking, but the whole occupation sequence could be of seventh-century date.

Pennyland

County:	Buckinghamshire
NGR:	SP 862 411
Main reference:	Williams 1993

Zooarchaeology:	J.M. Holmes in Williams 1993
Archaeobotany:	M. Jones in Williams 1993
Elevation:	80 m AOD
Geology:	Gravel spur projecting from a boulder clay ridge
Summary:	In the earliest phase, a scatter of *Grubenhäuser* and pits are datable to the Early Saxon period – in particular, perhaps, to the sixth century – by annular loomweights, an iron spearhead believed to be of sixth- to seventh-century type, and the occurrence of stamped and incised decoration on potsherds. An intermediate phase of occupation – consisting of *Grubenhäuser*, post-built structures, ditched enclosures and droveways – lacks both decorated sherds and Ipswich Ware, but includes a bone comb of mid- to late-sixth-century type, and the site morphology would be consistent with a seventh-century or later date. In the third phase, another scatter of *Grubenhäuser*, four small 'four-post structures', and a well are dated to the seventh to ninth centuries by the presence of small amounts of Ipswich Ware (*c.* AD 725–850) and a bone comb of late seventh- to eighth-century type.

Pitstone

County:	Buckinghamshire
NGR:	SP 9382 1507
Main reference:	Phillips 2005
Zooarchaeology:	E. Hambleton in Phillips 2005
Archaeobotany:	J. Robinson in Phillips 2005
Elevation:	120 m AOD
Geology:	Chalky head deposits above Lower Chalk bedrock, at the foot of the Chilterns scarp
Summary:	*Grubenhäuser* and pits, plus refuse in the upper fill of a Roman ditch. The pottery is mostly sand-tempered and undecorated, and is generally consistent with a fifth- to sixth-century date.

Ramsbury (High Street)

County:	Wiltshire
NGR:	SU 272 715
Main reference:	Haslam *et al.* 1980
Zooarchaeology:	J. Coy in Haslam *et al.* 1980
Archaeobotany:	n/a
Elevation:	111 m AOD
Geology:	River gravels and clay-with-flints in the Kennet valley
Summary:	Iron smelting site, with faunal remains from successive phases of activity, dated to the eighth and ninth centuries. Dating is based upon radiocarbon determinations from charcoal and the presence of a small number of datable artefacts such as strap ends.

Redcastle Furze, Thetford

County:	Norfolk
NGR:	TL 8615 8305
Main reference:	Andrews 1995
Zooarchaeology:	T. Wilson in Andrews 1995
Archaeobotany:	P. Murphy in Andrews 1995
Elevation:	14 m AOD
Geology:	Sands and gravels
Summary:	*Grubenhäuser*, pits and ditches represent an initial phase, followed by two shallow ditches of a subsequent phase. The first phase lacks Ipswich Ware but the rectilinear ditch system would be consistent with a seventh-century date. The second phase is dated by Ipswich Ware to *c.* AD 725–850.

Rycote (Site 30)

County:	Oxfordshire
NGR:	SP 6609 0495
Main reference:	Taylor & Ford 2004
Zooarchaeology:	S. Hamilton-Dyer in Taylor & Ford 2004
Archaeobotany:	M. Robinson in Taylor & Ford 2004
Elevation:	85 m AOD
Geology:	Lower Greensand and gravel
Summary:	Two *Grubenhäuser*, four parallel ditches, five pits and two postholes are thought to represent at least two phases of Anglo-Saxon activity. The fill of one *Grubenhaus* contained a ceramic assemblage dominated by sandy wares, some sherds bearing incised and/ or stamped decoration, and a bone radiocarbon-dated to cal. AD 437–637. The fill of a later *Grubenhaus*, which cut one of the ditches, contained a ceramic assemblage dominated by organic-tempered sherds, none of which were decorated, and a tooth radiocarbon-dated to cal. AD 646–758. The other features are less closely datable.

St Aldate's, Oxford: 79-80

County:	Oxfordshire
NGR:	SP 51400 05825
Main reference:	Durham 1977
Zooarchaeology:	B.J. Marples in Durham 1977
Archaeobotany:	n/a
Elevation:	57 m AOD
Geology:	Seasonally flooded islands of alluvial clay on the Upper Thames floodplain
Summary:	Complex sequence including revetments and a clay bank which may represent a causeway. Thermoluminescence dating of pottery from this phase returned a date range of AD 631–779.

St Aldate's, Oxford: 89-91 (Trill Mill Stream)

County:	Oxfordshire
NGR:	SP 51365 05910

Main reference: Dodd 2003
Zooarchaeology: B. Wilson in Dodd 2003
Archaeobotany: M. Robinson in Dodd 2003
Elevation: 53 m AOD
Geology: Seasonally flooded islands of alluvial clay on the Upper Thames floodplain
Summary: Wattle revetments radiocarbon-dated to cal. AD 660–990 and 680–1020. This phase is succeeded by a Late Saxon phase dated to the tenth century by pottery and other radiocarbon dates.

St Aldate's, Oxford: BT Tunnel
County: Oxfordshire
NGR: SP 51430 05645
Main reference: Dodd 2003
Zooarchaeology: n/a
Archaeobotany: M. Robinson in Dodd 2003
Elevation: 53 m AOD
Geology: Seasonally flooded islands of alluvial clay on the Upper Thames floodplain
Summary: Cobbled surface, possibly of Early Saxon date, whose dating is based upon Robinson's palaeoenvironmental sequence for the development of the floodplain, and also upon the felling date (AD 577–619) of a plank found in the overlying silt. A putative bridge trestle has been radiocarbon-dated to cal. AD 660–900.

Scole (Oakley/Scole, Waveney valley)
County: Suffolk
NGR: TM 147 786
Main reference: Ashwin & Tester 2014
Zooarchaeology: n/a
Archaeobotany: n/a
Elevation: 24 m AOD
Geology: Alluvium, sands and gravels in the valley; boulder clay plateaux to the north and south
Summary: Two parallel alignments of oak piles along the edge of a palaeochannel in the Waveney valley, interpreted as a revetment or causeway. Radiocarbon dates from two piles fall within the fifth to seventh centuries, while a third determination is of fourth- to sixth-century date. The excavated site falls partially within Norfolk, but the relevant data here are from Suffolk.

Sedgeford (Chalkpit Field)
County: Norfolk
NGR: TF 711 363
Main reference: Davies 2008; Bates 1991
Zooarchaeology: K. Poole in Davies 2008
Archaeobotany: V. Fryer in Davies 2008
Elevation: 22 m AOD
Geology: Sand and gravel overlying degraded chalk
Summary: An earlier phase is dated by Ipswich Ware to c. AD 725–850. A subsequent phase

in which Ipswich and Thetford ware 'overlap' is dated to the mid- to late-ninth century. Each phase includes ditches and an oven.

Shakenoak
County: Oxfordshire
NGR: SP 374 138
Main reference: Brodribb et al. 2005
Zooarchaeology: Brodribb et al. 2005
Archaeobotany: n/a
Elevation: 109 m AOD
Geology: Cotswold limestone, near junction with clays and river gravels in the Upper Thames valley
Summary: Poorly-understood post-Roman occupation at a Roman villa site, represented by artefactual and faunal material in the fill of a boundary ditch, originally (but perhaps tenuously) dated to the seventh and possibly eighth centuries.

Slough House Farm
County: Essex
NGR: TL 873 091
Main reference: Wallis & Waughman 1998
Zooarchaeology: n/a
Archaeobotany: P. Murphy in Wallis & Waughman 1998
Elevation: 10 m AOD
Geology: River terrace gravels
Summary: Pits, wells and postholes. Dendrochronology of timbers from the wells suggests that they post-date AD 599 and AD 602/3 respectively. One potsherd bears incised chevron decoration. Botanical samples from the wells are provisionally considered to be of seventh-century date.

Spong Hill
County: Norfolk
NGR: TF 981 195
Main reference: Rickett 1995
Zooarchaeology: J.M. Bond in Rickett 1995
Archaeobotany: P. Murphy in Rickett 1995
Elevation: 35 m AOD
Geology: Gravels
Summary: Six or seven *Grubenhäuser*, four 'posthole groups', pits, ditches and hollows. The ceramic assemblage, around 24% of which bore incised or stamped decoration, and with organic inclusions constituting the largest fabric group, could be consistent with occupation centred on the later sixth century.

Spring Road, Abingdon
County: Oxfordshire
NGR: SU 4875 9755
Main reference: Allen & Kamash 2008
Zooarchaeology: B. M. Charles in Allen & Kamash 2008

Archaeobotany: M. Robinson in Allen & Kamash 2008
Elevation: 59m AOD
Geology: Second gravel terrace of the Upper Thames
Summary: *Grubenhäuser*, postholes and ditches are assigned to a sixth- to seventh-century phase based upon the ceramic evidence, which includes sandy and organic-tempered fabrics. Incised and (especially) stamped decoration is here taken as an indicator of sixth-century date.

Stansted Airport
County: Essex
NGR: TL 523 224
Main reference: Havis & Brooks 2004
Zooarchaeology: n/a
Archaeobotany: P. Murphy in Havis & Brooks 2004
Elevation: 100 m AOD
Geology: Boulder clay plateau
Summary: A pit and shallow depression, whose fills are assigned a sixth- to seventh-century date by the ceramic assemblage, which is dominated by organic-tempered sherds and includes some incised decoration.

Stansted Carpark
County: Essex
NGR: TL 524 227
Main reference: Cooke *et al.* 2008
Zooarchaeology: n/a
Archaeobotany: n/a
Elevation: 100 m AOD
Geology: Boulder clay plateau
Summary: A hearth or oven with *in situ* burning. Charcoal from the feature has been radiocarbon-dated to cal. AD 680–890.

Stonea Grange
County: Cambridgeshire
NGR: TL 449 937
Main reference: Jackson & Potter 1996
Zooarchaeology: S. Stallibrass in Jackson & Potter 1996
Archaeobotany: M. van der Veen in Jackson & Potter 1996
Elevation: 4 m AOD
Geology: Sand and gravel deposits constituting the island of Stonea in the peat fens
Summary: Post-Roman phase of a Roman settlement, comprising at least four post-built structures along the line of a Roman road, plus enclosure ditches and pits. The ceramic assemblage, including some stamped sherds, is given a broad fifth- to seventh-century date range in the excavation report.

The Strood
County: Essex
NGR: TM 0150 1485
Main reference: Crummy *et al.* 1982
Zooarchaeology: n/a
Archaeobotany: n/a

Elevation: 3 m AOD
Geology: Alluvium overlying London Clay
Summary: Oak piles of an artificial causeway. Dendrochronology of the timbers suggests felling dates *c.* AD 684–702.

Sutton Courtenay (Drayton Road)
County: Oxfordshire
NGR: SU 490 936
Main reference: Hamerow *et al.* 2007
Zooarchaeology: B.M. Charles in Hamerow *et al.* 2007
Archaeobotany: M. Robinson in Hamerow *et al.* 2007
Elevation: 55 m AOD
Geology: Second terrace gravel of the Upper Thames
Summary: Multi-period site with extensive Anglo-Saxon occupation. Aerial survey and recent excavations have discovered a high-status complex of large timber halls dating from the seventh century.

Taplow Court
County: Buckinghamshire
NGR: SU 907 823
Main reference: Allen *et al.* 2009
Zooarchaeology: E.-J. Evans in Allen *et al.* 2009
Archaeobotany: M. Robinson in Allen *et al.* 2009
Elevation: 65 m AOD
Geology: Black Park terrace gravel over Upper Chalk
Summary: Reoccupation of an Iron Age hillfort, represented by domestic debris in surrounding ditches. Charred cereal remains have been radiocarbon-dated to cal. AD 770–980 and 670–780, and further seventh- to ninth-century radiocarbon dates were also obtained from charcoal found in postholes.

Terrington St. Clement
County: Norfolk
NGR: TF 5386 1800
Main reference: Crowson *et al.* 2005
Zooarchaeology: P. Baker in Crowson *et al.* 2005
Archaeobotany: P. Murphy in Crowson *et al.* 2005
Elevation: 3 m AOD
Geology: Silt roddon (a raised fenland island)
Summary: Pits, ditches, gullies, and a 'pond' dated by Ipswich Ware to *c.* AD 725–850. The larger ditches are thought to have served drainage purposes, while smaller ditches demarcated field boundaries and droveways.

Two Mile Bottom
County: Norfolk
NGR: TL 852 868
Main reference: Bates & Lyons 2003
Zooarchaeology: n/a
Archaeobotany: P. Murphy in Bates & Lyons 2003
Elevation: 10 m AOD
Geology: Sands and gravels

Summary: Pits, ditches and a possible *Grubenhaus*. A fifth- to early eighth-century date may be suggested by the lack of Ipswich Ware in the ceramic assemblage, along with the occurrence of Early Saxon indicators such as sherds with scored decoration.

Walton Lodge
County:	Buckinghamshire
NGR:	SP 8238 1324
Main reference:	Dalwood *et al.* 1989
Zooarchaeology:	P. Sadler in Dalwood *et al.* 1989
Archaeobotany:	J. Giorgi in Dalwood *et al.* 1989
Elevation:	84 m AOD
Geology:	Portland Beds, of soft limestone and sands
Summary:	Post-built structures, postholes, gullies, small pits and layers. A single sherd of Ipswich Ware suggests occupation in or by the early eighth century.

Walton Orchard
County:	Buckinghamshire
NGR:	SP 8238 1328
Main reference:	Ford & Howell 2004
Zooarchaeology:	S. Hamilton-Dyer in Ford & Howell 2004
Archaeobotany:	J. Letts in Ford & Howell 2004
Elevation:	84 m AOD
Geology:	Portland Beds, of soft limestone and sands
Summary:	Eight posthole buildings, one beamslot building, one *Grubenhaus*, fences, pits, gullies and hearths representing several phases of Anglo-Saxon activity. The ceramic assemblage includes two stamped sherds, two sherds of possible Ipswich Ware, and three of possible Maxey-type ware. The sparseness and likely residuality or redeposition of much of the pottery – and associated faunal and botanical material – precludes any closer chronological definition at this site. However, the axial arrangement of the buildings and beamslot construction of one of them suggests that the activity occurred in the seventh century or later.

Walton Road Stores
County:	Buckinghamshire
NGR:	SP 8245 1335
Main reference:	Bonner 1997
Zooarchaeology:	Sadler 1998
Archaeobotany:	Robinson 1997
Elevation:	85 m AOD
Geology:	Degraded clayey limestone with patches of sand
Summary:	Seven *Grubenhäuser*, two possible post-built structures, gullies, pits and fence lines. Available typescripts were in an unfinished state, but suggest that the distribution of different ceramic fabrics

reveals a division in the *Grubenhäuser*, between those of fifth- to sixth-century date and those of seventh- to eighth-century date.

Walton Street (82-84 Walton St, Aylesbury)
County:	Buckinghamshire
NGR:	SP 8225 1325
Main reference:	Stone 2009
Zooarchaeology:	J. Morris in Stone 2009
Archaeobotany:	A. Livarda in Stone 2009
Elevation:	80 m AOD
Geology:	Portland Beds, of soft limestone and sands
Summary:	Pits and gullies. The ceramic assemblage, including a single Ipswich Ware sherd and a relatively high proportion of organic-tempered sherds, is thought to be consistent with a late sixth- to eighth-century date.

Walton Vicarage
County:	Buckinghamshire
NGR:	SP 822 132
Main reference:	Farley 1976
Zooarchaeology:	B. Noddle in Farley 1976
Archaeobotany:	M. Monk in Farley 1976
Elevation:	84 m AOD
Geology:	Portland Beds, of soft limestone and sands
Summary:	Five *Grubenhäuser*, three post-built structures, gullies and pits, not closely datable within the fifth- to ninth-century period.

Wantage (Mill Street)
County:	Oxfordshire
NGR:	SU 3958 8814
Main reference:	Holbrook & Thomas 1996
Zooarchaeology:	M. Maltby in Holbrook & Thomas 1996
Archaeobotany:	n/a
Elevation:	88 m AOD
Geology:	Near the interface of clay vales and calcareous head deposits at the scarp of the Lambourn Downs
Summary:	A Romano-British site overlain by a series of ditched enclosures; a silt layer, interpreted as a redeposited midden, overlies the ditches. The ditch fills and silt horizon are dated by the excavator to the fifth to seventh centuries. Handmade Anglo-Saxon pottery from these contexts is dominated by organic-tempered wares, and a single sherd bears stamped decoration. Morphologically, the enclosures are suggestive of a seventh-century or later date.

Waterbeach (Denny End)
County:	Cambridgeshire
NGR:	TL 4935 6572
Main reference:	Mortimer 1996

Zooarchaeology:	E. Yannouli in Mortimer 1996
Archaeobotany:	C. Stevens in Mortimer 1996
Elevation:	6 m AOD
Geology:	Second terrace river gravels and sands
Summary:	A *Grubenhaus*, pits, postholes and hollows are assigned an Early Saxon date by the excavator, in view of the absence of Ipswich Ware and the presence of a plain bowl and some decorated sherds which are thought to have been made in the fifth to sixth centuries.

Wavendon Gate

County:	Buckinghamshire
NGR:	SP 903 369
Main reference:	Williams *et al.* 1996
Zooarchaeology:	K. Dobney & D. Jacques in Williams *et al.* 1996
Archaeobotany:	J. Letts in Williams *et al.* 1996
Elevation:	88 m AOD
Geology:	Sand and gravel deposit upon a ridge of boulder clay
Summary:	Ephemeral Anglo-Saxon occupation on the site of a Roman settlement, including pits and postholes. The sixth-century date favoured in the excavation report is apparently based upon a single stamped sherd thought to be of sixth-century type, but the remainder of the ceramic assemblage (dominated by sandy fabrics, with a relatively low proportion of organic-tempered wares) would also be consistent with this chronology.

Westbury-by-Shenley

County:	Buckinghamshire
NGR:	SP 829 356
Main reference:	Ivens *et al.* 1995
Zooarchaeology:	n/a
Archaeobotany:	J. Letts in Ivens *et al.* 1995
Elevation:	110 m AOD
Geology:	Oxford Clay and boulder clay
Summary:	Two wells and a pit. Timber and a hazel hurdle from the wells returned radiocarbon dates spanning the mid-seventh to mid-eighth centuries.

Whitehouse Road, Ipswich (Bramford)

County:	Suffolk
NGR:	TM 138 470
Main reference:	Caruth 1996
Zooarchaeology:	n/a
Archaeobotany:	Fryer & Murphy 1996
Elevation:	30 m AOD
Geology:	Clay, sands and gravels
Summary:	Pits, posthole buildings and a cemetery, all within enclosure ditches. Only interim reports were available at the time of writing, but occupation is believed to begin around the eighth century (dated by two *sceattas*) and continue into the ninth or tenth century.

Wickhams Field

County:	Berkshire
NGR:	SU 6750 6970
Main reference:	Crockett 1996
Zooarchaeology:	S. Hamilton-Dyer in Crockett 1996
Archaeobotany:	R. Scaife in Crockett 1996
Elevation:	43 m AOD
Geology:	The pits are on the higher sandy ground of the Reading Beds, the wells are on lower valley gravels
Summary:	Three pits and two timber-lined wells. A well timber has been radiocarbon-dated to cal. AD 650–870; the other well was considered contemporary on morphological grounds. The small ceramic assemblage from the pits, consisting of organic-tempered sherds, would be consistent with a similar seventh-century or later date.

Willingham (High Street)

County:	Cambridgeshire
NGR:	TL 4040 7037
Main reference:	Fletcher 2008
Zooarchaeology:	C. Faine in Fletcher 2008
Archaeobotany:	V. Fryer in Fletcher 2008
Elevation:	6 m AOD
Geology:	First terrace sand and gravel deposits and Ampthill Clays
Summary:	Post-built structure, other postholes, pits, and possibly a ditch, dated to the eighth and ninth centuries by Ipswich Ware and North French Blackware.

Wilton (Wilton Autos, West St)

County:	Wiltshire
NGR:	SU 09420 31370
Main reference:	De'Athe 2012
Zooarchaeology:	Grimm 2008
Archaeobotany:	R. Pelling in De'Athe 2012
Elevation:	55 m AOD
Geology:	Gravel spur between the Nadder and the Wylye rivers
Summary:	One large *Grubenhaus*, plus associated interior features such as postholes and hearths. These contained annular loomweights and two sherds of organic-tempered pottery. The artefactual evidence is slim, but the relatively large size of the *Grubenhaus* might support a seventh-century date.

Wittering (Bonemills Farm)

County:	Cambridgeshire
NGR:	TF 0475 0153
Main reference:	Wall 2011
Zooarchaeology:	n/a
Archaeobotany:	A. Clapham in Wall 2011
Elevation:	50 m AOD
Geology:	Ironstone outcrop
Summary:	Iron smelting site with furnaces, hearths and pits. Charcoal samples from the

features have been radiocarbon-dated to cal. AD 575–875, 680–905, and 920–950. Environmental samples were dominated by charcoal, but semi-quantitative data for charred grains and seeds were available for three of the samples.

Wolverton Turn
County: Buckinghamshire
NGR: SP 8025 4066
Main reference: Preston 2007
Zooarchaeology: N. Sykes in Preston 2007
Archaeobotany: M Robinson & J. Letts in Preston 2007
Elevation: 79 m AOD
Geology: Mainly silty clay, with occasional outcropping limestone
Summary: A large sub-rectangular enclosure. A ceramic assemblage including Ipswich Ware, plus bone radiocarbon-dated to cal. AD 690–890, suggest that the enclosure ditches were associated with eighth- to ninth-century activity. A kiln or oven nearby may plausibly be associated with the same phase of activity. A *Grubenhaus* to the north, however, contained bone radiocarbon-dated to cal. AD 430–600.

Uley (West Hill)
County: Gloucestershire
NGR: ST 789 997
Main reference: Woodward & Leach 1993
Zooarchaeology: B. Levitan in Woodward & Leach 1993
Archaeobotany: n/a
Elevation: 246 m AOD
Geology: Spur of Cotswolds escarpment, bearing thin soils
Summary: Post-Roman sequence at a Romano-British shrine, including possible ecclesiastical structures and a faunal assemblage. Handmade Anglo-Saxon pottery in limestone and organic-tempered fabrics suggest a fifth- to seventh-century occupation sequence.

Walpole St. Andrew
County: Norfolk
NGR: TF 4874 1600
Main reference: Crowson *et al.* 2005
Zooarchaeology: P. Baker in Crowson *et al.* 2005
Archaeobotany: P. Murphy in Crowson *et al.* 2005
Elevation: 3 m AOD
Geology: Silt roddon (a raised fenland island)
Summary: Pits, ditches and gullies dated by Ipswich Ware to *c.* AD 725–850.

West Stow
County: Suffolk
NGR: TL 7970 7135
Main reference: West 1985
Zooarchaeology: Crabtree 1989; 2012

Archaeobotany: P. Murphy in West 1985
Elevation: 16 m AOD
Geology: Sands and gravels, with patches of brickearth, in the Lark valley
Summary: *Grubenhäuser*, post-built structures, postholes, pits and hollows. The *Grubenhäuser* are divided into three phases on stratigraphic and artefactual grounds, corresponding broadly to the fifth to sixth, sixth to seventh, and seventh to early eighth centuries respectively. Ditch systems at the site belong to the latter phase.

Wicken Bonhunt
County: Essex
NGR: TL 511 335
Main reference: Wade 1980
Zooarchaeology: Crabtree 2012
Archaeobotany: A.K.G. Jones 1975
Elevation: 65 m AOD
Geology: Gravel and sandy clay
Summary: Ditch systems, wells, a possible mill-leat, and post-built structures dated to the eighth and ninth centuries by Ipswich Ware, imported French wares, and timbers (radiocarbon-dated to cal. AD 780–880). Pits and ditches of a preceding phase, associated with handmade sand-tempered pottery, are presumed to date from around the seventh and early eighth centuries.

Witton
County: Norfolk
NGR: TG 336 320
Main reference: Lawson 1983
Zooarchaeology: n/a
Archaeobotany: A.K.G. Jones in Lawson 1983
Elevation: 13 m AOD
Geology: Sands and gravels
Summary: *Grubenhäuser*, a post-built structure, pits and hearths. Pottery bearing stamping, corrugation, rustication, and grooving is considered to support a fifth- to sixth-century date.

Worton
County: Oxfordshire
NGR: SP 460 112
Main reference: Hey 2004
Zooarchaeology: J. Mulville & K. Ayres in Hey 2004
Archaeobotany: M. Robinson in Hey 2004
Elevation: 64 m AOD
Geology: Second gravel terrace of the Upper Thames
Summary: Three *Grubenhäuser*, a posthole and possible pit are assigned to an Early Saxon phase, whose ceramic assemblage – dominated by organic-tempered sherds

and including stamped decoration – may suggest a late sixth- to seventh-century date. A subsequent Mid Saxon phase is represented by a post-in-trench building; charred grain from a post-pipe in a wall-trench has been radiocarbon-dated to cal. AD 640–880.

Wymondham (Browick Road)

County:	Norfolk
NGR:	TG 124 015
Main reference:	Ames 2005
Zooarchaeology:	n/a
Archaeobotany:	Anonymous report in Ames 2005
Elevation:	41 m AOD
Geology:	Boulder clay plateau, thinly covered by windblown sand
Summary:	A hearth or oven and a probable *Grubenhaus* containing pottery of 'Early/Mid Saxon' (sixth- to ninth-century) date.

Yarnton

County:	Oxfordshire
NGR:	SP 475 113
Main reference:	Hey 2004
Zooarchaeology:	J. Mulville & K. Ayres in Hey 2004
Archaeobotany:	C. Stevens & M. Robinson in Hey 2004
Elevation:	6 m AOD
Geology:	Second gravel terrace of the Upper Thames
Summary:	A complex settlement sequence at a multi-phase site, with chronological difficulties compounded by a lack of diagnostic ceramics. Stratigraphic relationships, and an extensive radiocarbon-dating programme, suggest the following phasing (some of which is, nonetheless, circumstantial). Four *Grubenhäuser* and a small post-built structure belong to an Early Saxon phase, dated to the late fifth or early sixth century. A larger *Grubenhaus* with a pit and postholes is probably of seventh-century date. From the late seventh or early eighth century, the character of the settlement was transformed: post-built structures, *Grubenhäuser*, ditched enclosures, paddocks and a possible trackway, pits, fence lines, and specialist buildings interpreted as granaries and a 'fowl house' were constructed. In the later eighth and ninth centuries, the ditches were recut on a larger, deeper scale. New post-built structures were created or rebuilt, and further pits or wells were dug.

Bibliography

Abrams, J. & Ingham, D. (2008). *Farming on the Edge: Archaeological Evidence from the Clay Uplands to the West of Cambridge*. (Bedford: Albion Archaeology).

Allen, T., Barton, N. & Brown, A. (1995). *Lithics and Landscape: Archaeological Discoveries on the Thames Water pipeline at Gatehampton Farm, Goring, Oxfordshire, 1985–92*. (Oxford: Oxford University Committee for Archaeology).

Allen, T., Cramp, K., Lamdin-Whymark, H. & Webley, L. (2010). *Castle Hill and its Landscape; Archaeological Investigations at the Wittenhams, Oxfordshire*. (Oxford: Oxford Archaeology).

Allen, T., Hayden, C. & Lamdin-Whymark, H. (2009). *From Bronze Age Enclosure to Anglo-Saxon Settlement. Archaeological Excavations at Taplow hillfort, Buckinghamshire, 1999–2005*. (Oxford: Oxford Archaeology).

Allen, T. & Kamash, Z. (2008). *Saved From the Grave: Neolithic to Saxon Discoveries at Spring Road Municipal Cemetery, Abingdon, Oxfordshire, 1990–2000*. (Oxford: Oxford Archaeology).

Ames, J. (2005). *An Archaeological Evaluation at Browick Road, Wymondham, Norfolk*. (unpublished report, Norfolk Archaeological Unit).

Amorosi, T., Woollett, J., Perdikaris, S. & McGovern, T. (1996). Regional zooarchaeology and global change: problems and potentials. *World Archaeology* 28(1), 126–157.

Andrews, P. (1995). *Excavations at Redcastle Furze, Thetford, 1988–9*. East Anglian Archaeology 72. (Gressenhall: Norfolk Museums Service).

Andrews, P., Biddulph, E., Hardy, A. & Brown, R. (2011). *Settling the Ebbsfleet Valley. High Speed 1 Excavations at Springhead and Northfleet, Kent. The Late Iron Age, Roman, Saxon, and Medieval Landscape. Volume 1: The Sites*. (Oxford & Salisbury: Oxford Wessex Archaeology).

Andrews, P., Mepham, L., Schuster, J. & Stevens, C. J. (2011). *Settling the Ebbsfleet Valley. High Speed 1 Excavations at Springhead and Northfleet, Kent. The Late Iron Age, Roman, Saxon, and Medieval Landscape. Volume 4: Saxon and Later Finds and Environmental Reports*. (Oxford & Salisbury: Oxford Wessex Archaeology).

Arnold, C. J. & Wardle, P. (1981). Early medieval settlement patterns in England. *Medieval Archaeology* 25, 145–149.

Arthur, P., Fiorentino, G. & Grasso, A. M. (2012). Roads to recovery: an investigation of early medieval agrarian strategies in Byzantine Italy in and around the eighth century. *Antiquity* 86(332), 444–455.

Ashwin, T. & Tester, A. (2014). *A Romano-British Settlement in the Waveney Valley: Excavations at Scole, 1993–4*. East Anglian Archaeology 152. (Gressenhall: Norfolk Historic Environment Service).

Atkins, R. & Connor, A. (2010). *Farmers and Ironsmiths: Prehistoric, Roman and Anglo-Saxon Settlements beside Brandon Road, Thetford, Norfolk*. (Bar Hill: Oxford Archaeology East).

Atkinson, M. & Preston, S. (1998). The Late Iron Age and Roman settlement at Elms Farm, Heybridge, Essex, Excavations 1993–5: An interim report. *Britannia* 29, 85–110.

Audouy, M. & Chapman, A. (2009). *Raunds: the Origin and Growth of a Midland Village, AD 450–1500*. (Oxford: Oxbow Books).

Bakels, C. (2005). Crops produced in the southern Netherlands and northern France during the early medieval period: a comparison. *Vegetation History and Archaeobotany* 14, 394–399.

Banham, D. (2004). *Food and Drink in Anglo-Saxon England*. (Stroud: Tempus).

Banham, D. (2010). 'In the Sweat of thy Brow Shalt thou eat Bread': cereals and cereal production in the Anglo-Saxon landscape. In: Higham, N. J. & Ryan, M. J. eds. *The Landscape Archaeology of Anglo-Saxon England*. (Woodbridge: Boydell Press), 175–192.

Banham, D. & Faith, R. (2014). *Anglo-Saxon Farms and Farming*. (Oxford: Oxford University Press).

Barclay, A., Lambrick, G., Moore, J. & Robinson, M. (2003). *Lines in the Landscape. Cursus monuments in the Upper Thames Valley: excavations at the Drayton and Lechlade cursuses*. (Oxford: Oxford Archaeology).

Bartosiewicz, L., Van Neer, W. & Lentacker, A. (1993). Metapodial asymmetry in draft cattle. *International Journal of Osteoarchaeology* 3(2), 69–75.

Bartosiewicz, L., Van Neer, W. & Lentacker, A. (1997). *Draught Cattle: their Osteological Identification and History*. (Tervuren: Musée Royal de l'Afrique Centrale).

Bateman, C., Enright, D. & Oakey, N. (2003). Prehistoric to Anglo-Saxon settlements to the rear of Sherborne House, Lechlade: excavations in 1997. *Transactions of the Bristol and Gloucestershire Archaeological Society* 121, 23–96.

Bates, S. (1991). *Summary Report of Excavations at Sedgeford, Norfolk, June - July 1991*. (unpublished report, Norfolk Archaeological Unit).

Bates, S. & Lyons, A. (2003). *The Excavation of Romano-British Pottery Kilns at Ellingham, Postwick and Two Mile Bottom, Norfolk, 1995–7*. East Anglian Archaeology Occasional Paper 13 (Dereham: Norfolk Museums and Archaeological Service).

Beckett, J. V (1989). *A History of Laxton: England's Last Open-Field Village*. (Oxford: Blackwell).

Behre, K. E. (1992). The history of rye cultivation in Europe. *Vegetation History and Archaeobotany* 1, 141–156.

Bennett, P., Riddler, I. & Sparey-Green, C. (2010). *The Roman Watermills and Settlement at Ickham, Kent*. (Canterbury: Canterbury Archaeological Trust).

Blair, J. (2005). *The Church in Anglo-Saxon Society*. (Oxford: Oxford University Press).

Blair, J. (2013b). Grid-planning in Anglo-Saxon settlements: the short perch and the four-perch module. *Anglo-Saxon Studies in Archaeology and History* 18, 18–61.

Blair, J. (2013a). *The British Culture of Anglo-Saxon Settlement (H.M. Chadwick Memorial Lectures 24)*. (Cambridge: Department of Anglo-Saxon, Norse and Celtic, University of Cambridge).

Blinkhorn, P. (2012). *The Ipswich Ware Project: Ceramics, Trade and Society in Middle Saxon England*. (London: Medieval Pottery Research Group).

Boardman, S. & Jones, G. (1990). Experiments on the effects of charring on cereal plant components. *Journal of Archaeological Science* 17, 1–11.

Bonner, D. (1997). Untitled draft typescript concerning excavations at Walton, Aylesbury. (on file at Buckinghamshire Historic Environment Records).

Booth, P., Dodd, A., Robinson, M. & Smith, A. (2007). *The Thames Through Time. The Archaeology of the Gravel Terraces of the Upper and Middle Thames. The Early Historical Period: AD 1–1000*. Thames Through Time 3. (Oxford: Oxford Archaeology).

Booth, P., Evans, J. A. & Hiller, J. (2001). *Excavations in the Extramural Settlement of Roman Alchester, Oxfordshire, 1991*. (Oxford: Oxford Archaeology).

Boulter, S. (2005). *Handford Road, Ipswich (IPS 280), Archaeological Assessment Report (Volume I: Text)*. (unpublished report, Suffolk County Council Archaeological Service).

Boulter, S. (2008). *An Assessment of the Archaeology Recorded in New Phases 5, 6, 7(a & b), 9, 11 & 12 of Flixton Park Quarry*. (unpublished report, Suffolk County Council Archaeological Service).

Boulter, S. (2010). *Archaeological Assessment Report. Silver Birches, Hintlesham (HNS 027)*. (unpublished report, Suffolk County Council Archaeological Service).

Bowen, H. C. (1961). *Ancient Fields. A Tentative Analysis of Vanishing Earthworks and Landscapes*. (London: British Association for the Advancement of Science).

Bradley, R. (1978). Rescue Excavation in Dorchester-on-Thames 1972. *Oxoniensia* 43, 17–39.

Bradley, R. (2006). Bridging two cultures – commercial archaeology and the study of prehistoric Britain. *Antiquaries Journal* 86, 1–13.

Branigan, K. (1971). *Latimer. Belgic, Roman, Dark Age and Early Modern Farm*. (Bristol: Chess Valley Archaeological and Historical Society).

Brennan, N. & Hamerow, H. (2015). An Anglo-Saxon Great Hall complex at Sutton Courtenay/Drayton, Oxfordshire: a Royal centre of early Wessex? *Archaeological Journal* 172, 325–350.

Brodribb, A., Hands, A. & Walker, D. (2005). *The Roman Villa at Shakenoak Farm, Oxfordshire. Excavations 1960–1976*. British Archaeological Report 395. (Oxford: British Archaeological Reports).

Brown, T. & Foard, G. (1998). The Saxon landscape: a regional perspective. In: Everson, P. & Williamson, T. eds. *The Archaeology of Landscape*. (Manchester: Manchester University Press), 67–94.

Brunning, R. (2010). Taming the floodplain: river canalisation and causeway formation in the Middle Anglo-Saxon period at Glastonbury, Somerset. *Medieval Archaeology* 54, 319–329.

Büntgen, U., Tegel, W., Nicolussi, K., McCormick, M., Frank, D., Trouet, V., Kaplan, J. O., Herzig, F., Heussner, K. U., Wanner, H., Luterbacher, J. & Esper, J. (2011). 2500 years of european climate variability and human susceptibility. *Science* 331, 578–582.

Butzer, K. W. (1993). The classical tradition of agronomic science: perspectives on Carolingian agriculture and agronomy. In: Butzer, P. L. & Lohrmann, D. eds. *Science in Western and Eastern Civilization in Carolingian Times*. (Basel: Birkhäuser), 539–596.

Campbell, G. (1994). The preliminary archaeobotanical results from Anglo-Saxon West Cotton and Raunds. In: Rackham, J. ed. *Environment and Economy in Anglo-Saxon England*. Council for British Archaeology Research Report 89. (York: Council for British Archaeology), 65–82.

Campbell, G. (2012). *Assessment of Charred and Mineral-replaced Macroscopic Plant Remains from Excavation at Lyminge, Kent, 2008–10*. (unpublished report for University of Reading).

Cappers, R. T. J. & Neef, R. (2012). *Handbook of Plant Palaeoecology*. (Groningen: Barkhuis & Groningen University Library).

Caruth, J. (1996). Ipswich, Hewlett Packard plc, Whitehouse Industrial Estate. *Proceedings of the Suffolk Institute of Archaeology and History* 38(4), 476–477.

Caruth, J. (2006). *Consolidated Support Complex, RAF Lakenheath, ERL 116 and Family Support Complex, RAF Lakenheath ERL 139. A Report on the Archaeological Excavations, 2001–2005*. (unpublished report, Suffolk County Council Archaeological Service).

Caruth, J. (2008). Eye, Hartismere High School. *Proceedings of the Suffolk Institute of Archaeology and History* 41(4), 518–520.

Chambers, R. & McAdam, E. (2007). *Excavations at Barrow Hills, Radley, Oxfordshire, 1983–5*. (Oxford: Oxford Archaeology).

Chapman, A. (2010). *West Cotton, Raunds: a Study of Medieval Settlement Dynamics, AD 450–1450*. (Oxford: Oxbow Books).

Chapman, H. (2006). *Landscape Archaeology and GIS*. (Stroud: Tempus).

Clapham, A. R., Tutin, T. G. & Warburg, E. F. (1962). *Flora of the British Isles*. 2nd ed. (Cambridge: Cambridge University Press).

Clutton-Brock, J. (1976). The animal resources. In: Wilson, D. M. ed. *The Archaeology of Anglo-Saxon England*. (London: Methuen), 373–392.

Coleman, L., Hancocks, A. & Watts, M. (2006). *Excavations on the Wormington to Tirley Pipeline, 2000. Four Sites by the Carrant Brook and River Isbourne Gloucestershire and Worcestershire*. (Cirencester: Cotswold Archaeology).

Cooke, N., Brown, F. & Phillpots, C. (2008). *From Hunter Gatherers to Huntsmen. A History of the Stansted Landscape*. (Oxford & Salisbury: Framework Archaeology).

Crabtree, P. J. (1989). *West Stow, Suffolk: Early Anglo-Saxon Animal Husbandry*. (Ipswich: Suffolk County Council).

Crabtree, P. J. (1996). Production and consumption in an early complex society: animal use in Middle Saxon East Anglia. *World Archaeology* 28(1), 58–75.

Crabtree, P. J. (2007). Animals as material culture in Middle Saxon England: the zooarchaeological evidence for wool production at Brandon. In: Pluskowski, A. ed. *Breaking and Shaping Beastly Bodies. Animals as Material Culture in the Middle Ages*. (Oxford: Oxbow Books), 161–169.

Crabtree, P. J. (2010). Agricultural innovation and socio-economic change in early medieval Europe: evidence from Britain and France. *World Archaeology* 42(1), 122–136.

Crabtree, P. J. (2012). *Middle Saxon animal husbandry in East Anglia*. (Bury St Edmunds: Suffolk County Council Archaeological Service).

Crockett, A. (1996). Iron Age to Saxon settlement at Wickhams Field, near Reading, Berkshire: excavations on the site of the

M4 motorway service area. In: Andrews, P. & Crockett, A. eds. *Three Excavations along the Thames and its Tributaries, 1994: Neolithic to Saxon Settlement and Burial in the Thames, Colne, and Kennet Valleys.* Wessex Archaeology Report 10. (Salisbury: Wessex Archaeology), 113–170.

Crowson, A., Lane, T., Penn, K. & Trimble, D. (2005). *Anglo-Saxon Settlement on the Siltland of Eastern England.* Lincolnshire Archaeology and Heritage Reports Series 7 (Sleaford: Heritage Trust of Lincolnshire).

Crummy, P., Hillam, J. & Crossan, C. (1982). Mersea Island: the Anglo-Saxon causeway. *Essex Archaeology and History* 14, 77–86.

Dale, R., Maynard, D., Tyler, S. & Vaughan, T. (2010). Carved in stone: a late Iron Age and Roman cemetery and evidence for a Saxon minster, excavations near St Nicholas' church, Great Wakering 1998 and 2000. *Essex Society for Archaeology and History* 1, 194–231.

Dalwood, H., Dillon, J., Evans, J. & Hawkins, A. (1989). Excavations in Walton, Aylesbury, 1985–1986. *Records of Buckinghamshire* 31, 137–225.

Daniels, R. (2007). *Anglo-Saxon Hartlepool and the Foundations of English Christianity. An Archaeology of the Anglo-Saxon Monastery.* Tees Archaeology Monograph Series (Hartlepool: Tees Archaeology).

Dark, P. (2000). *The Environment of Britain in the First Millennium AD.* (London: Duckworth).

Davies, G. (2008). *An Archaeological Evaluation of the Middle-Late Anglo-Saxon Settlement at Chalkpit Field, Sedgeford, Northwest Norfolk.* (unpublished draft report, http://www.scribd.com/doc/3989245/CNEreport-draft - accessed 17/11/2011).

Day, S. P. (1991). Post-glacial vegetational history of the Oxford region. *New Phytologist* 119(3), 445–470.

De'Athe, R. (2008). *Wilton Autos, 41–43 West Street, Wilton, Wiltshire. Post-Excavation Assessment and Proposals for Analysis and Publication.* (unpublished report, Wessex Archaeology).

De'Athe, R. (2012). Early to middle Anglo-Saxon settlement, a lost medieval church rediscovered and an early post-medieval cemetery in Wilton. *Wiltshire Archaeological and Natural History Magazine* 105, 117–144.

Denison, S. (2001). Mercian watermill found near Welsh border. *British Archaeology* 57, 5.

Dobney, K., Jacques, D., Barrett, J. & Johnstone, C. (2007). *Farmers, Monks and Aristocrats: the Environmental Archaeology of Anglo-Saxon Flixborough.* (Oxford: Oxbow).

Dodd, A. ed. (2003). *Oxford Before the University: the Late Saxon and Norman Archaeology of the Thames Crossing, the Defences and the Town.* (Oxford: Oxford Archaeology).

Dodwell, N., Lucy, S. & Tipper, J. (2004). Anglo-Saxons on the Cambridge Backs: the Criminology site settlement and King's Garden Hostel cemetery. *Proceedings of the Cambridge Antiquarian Society* 93, 95–124.

Durham, B. (1977). Archaeological investigations in St. Aldates, Oxford. *Oxoniensia* 42, 83–203.

Edwards, C. (2008). Saxon archaeology and medieval archaeology at Forbury House, Reading. *Berkshire Archaeological Journal* 77, 39–44.

Ellis, C.J. (2004). *A Prehistoric Ritual Complex at Eynesbury, Cambridgeshire. Excavation of a Multi-Period Site in the Great Ouse Valley, 2000–2001.* East Anglian Archaeology Occasional Paper 17. (Salisbury: Trust for Wessex Archaeology).

Evans, C., Mackay, D. & Webley, L. (2008). *Borderlands. The Archaeology of the Addenbrooke's Environs, South Cambridge.* CAU New Archaeologies of the Cambridge Region 1. (Cambridge: Cambridge Archaeological Unit).

Evans, D. H. & Loveluck, C. (2009). *Life and Economy at Early Medieval Flixborough, c. AD 600–1000: The Artefact Evidence.* (Oxford: Oxbow Books).

Faith, R. (1997). *The English Peasantry and the Growth of Lordship.* (London: Leicester University Press).

Faith, R. (2009). Forces and relations of production in early medieval England. *Journal of Agrarian Change* 9(1), 23–41.

Farley, M. (1976). Saxon and medieval Walton, Aylesbury: excavations 1973–4. *Records of Buckinghamshire* 20(2), 153–290.

Farley, M. & Jones, G. (2012). *Iron Age Ritual, a Hillfort and Evidence for a Minster at Aylesbury, Buckinghamshire.* (Oxford: Oxbow Books).

Fasham, P. J. & Whinney, R. J. B. (1991). *Archaeology and the M3. The Watching Brief, the Anglo-Saxon settlement at Abbots Worthy and retrospective sections.* Hampsire Field Club Monograph 7 (Winchester: Hampshire Field Club).

Fenton, A. (1978). *The Northern Isles: Orkney and Shetland.* (Edinburgh: John Donald).

Fischer, T. (1990). *Das Umland des römischen Regensburg.* (Munich: C.H. Beck).

Fleming, R. (2010). *Britain after Rome. The Fall and Rise, 400–1070.* Anglo-Saxon Britian 2(1) (London: Allen Lane).

Fletcher, J. & Tapper, M. (1984). Medieval artefacts and structures dated by dendrochronology. *Medieval Archaeology* 28, 112–132.

Fletcher, T. (2008). *Anglo-Saxon Settlement and Medieval Pits at 1 High Street, Willingham, Cambridgeshire.* (unpublished report, Cambridgeshire Archaeology Archaeological Field Unit).

Ford, S. (2002). *Charnham Lane, Hungerford, Berkshire; Archaeological Investigations 1988–1997.* (Reading: Thames Valley Archaeological Services).

Ford, S. & Howell, I. J. (2004). Saxon and Bronze Age settlement at the Orchard Site, Walton Road, Walton, Aylesbury, 1994. In: Ford, S., Howell, I. J., & Taylor, K. eds. *The Archaeology of the Aylesbury-Chalgrove Gas Pipeline and The Orchard, Walton Road, Aylesbury.* (Reading: Thames Valley Archaeological Services), 61–88.

Foreman, S., Hiller, J. & Petts, D. (2002). *Gathering the People, Settling the Land. The Archaeology of a Middle Thames Landscape: Anglo-Saxon to Post-medieval.* (Oxford: Oxford Archaeology).

Fowler, P. J. (2002). *Farming in the First Millennium AD: British Agriculture Between Julius Caesar and William the Conqueror.* (Cambridge: Cambridge University Press).

Fryer, V. (2008). *An Assessment of the Charred Plant Macrofossils and Other Remains from Eye, Suffolk (EYE 083).* (unpublished manuscript from J. Caruth, Suffolk County Council Archaeological Service).

Fryer, V. & Murphy, P. (1996). *Macrobotanical and Other Remains from Whitehouse Industrial Estate, Ipswich, Suffolk (IPS 247): An Assessment.* (unpublished manuscript from J. Caruth, Suffolk County Council Archaeological Service).

Gaimster, M., Haith, C. & Bradley, J. (1998). Medieval Britain and Ireland in 1997. *Medieval Archaeology* 42, 107–190.

Gardiner, M. (2011). Late Saxon settlements. In: Hamerow, H., Hinton, D. A., & Crawford, S. eds. *The Oxford Handbook of Anglo-Saxon Archaeology.* (Oxford: Oxford University Press), 198–217.

Gardiner, M. (2012). Stacks, barns and granaries in early and High Medieval England: crop storage and its implications.

In: Quirós Castillo, J. A. ed. *Horrea, Silos and Barns*. (Vitoria-Gasteiz: University of the Basque Country), 23–38.

Garrow, D., Lucy, S. & Gibson, D. (2006). *Excavations at Kilverstone, Norfolk: an Episodic Landscape History*. East Anglian Archaeology 113. (Cambridge: Cambridge Archaeological Unit).

Gibson, C. (2003). An Anglo-Saxon settlement at Godmanchester, Cambridgeshire. *Anglo-Saxon Studies in Archaeology and History* 12, 137–217.

Gray, H. L. (1915). *English Field Systems*. (Cambridge MA: Harvard University Press).

Gray, M. (1974). The Saxon settlement at New Wintles Farm, Eynsham. In: Rowley, T. ed. *Anglo-Saxon Settlement and Landscape: Papers Presented to a Symposium, Oxford 1973*. British Archaeological Report 6 (Oxford: British Archaeological Reports), 51–55.

Green, F. (1981). Iron Age, Roman and Saxon crops: the archaeological evidence from Wessex. In: Jones, M. & Dimbleby, G. eds. *The Environment of Man: the Iron Age to the Anglo-Saxon Period*. British Archaeological Report 87. (Oxford: British Archaeological Reports), 129–153.

Green, F. (1982). Problems of interpreting differentially preserved plant remains from excavations of medieval urban sites. In: Hall, A. & Kenward, H. eds. *Environmental Archaeology in the Urban Context*. Council for British Archaeology Research Report 43. (London: Council for British Archaeology), 40–46.

Green, J. R. (1885). *The Making of England*. (London: Macmillan).

Grimm, J. M. (2008). Wilton: Animal Bone Report (60515). (unpublished report from author).

Hagen, A. (2006). *Anglo-Saxon Food and Drink. Production, Processing, Distribution and Consumption*. (Hockwold cum Wilton: Anglo-Saxon Books).

Hall, A. & Huntley, J.P. (2007). *A Review of the Evidence for Macrofossil Plant Remains from Archaeological Deposits in Northern England*. English Heritage Research Department Report Series 87/2007. (Portsmouth: English Heritage).

Hall, D. (1982). *Medieval Fields*. (Princes Risborough: Shire).

Hall, D. (2014). *The Open Fields of England*. (Oxford: Oxford University Press).

Hall, R. V. (2003). *Archaeological Evaluation of Land North of the Post Office, Church Terrace, Outwell, Norfolk*. (unpublished report, Archaeological Project Services).

Hambleton, E. (1999). *Animal Husbandry Regimes in Iron Age Britain. A Comparative Study of Faunal Assemblages from British Iron Age Sites*. British Archaeological Report 282. (Oxford: British Archaeological Reports).

Hamerow, H. (1991). Settlement mobility and the 'middle Saxon shift': Rural settlements and settlement patterns in Anglo-Saxon England. *Anglo-Saxon England* 20, 1–18.

Hamerow, H. (1993). *Excavations at Mucking. Volume 2: The Anglo-Saxon Settlement*. (London: English Heritage).

Hamerow, H. (1997). Migration Theory and the Anglo-Saxon 'identity crisis'. In: Chapman, J. & Hamerow, H. eds. *Migrations and Invasions in Archaeological Explanation*. British Archaeological Report S664 (Oxford: British Archaeological Reports), 33–44.

Hamerow, H. (2002). *Early Medieval Settlements: the Archaeology of Rural Communities in Northwest Europe, 400–900*. (Oxford: Oxford University Press).

Hamerow, H. (2012). *Rural Settlements and Society in Anglo-Saxon England*. (Oxford: Oxford University Press).

Hamerow, H., Hayden, C. & Hey, G. (2007). Anglo-Saxon and earlier settlement near Drayton Road, Sutton Courtenay, Berkshire. *Archaeological Journal* 164, 109–196.

Hancock, A. (2010). Excavation of a mid-Saxon settlement at Water Eaton, Bletchley, Milton Keynes. *Records of Buckinghamshire* 50, 5–24.

Hardy, A., Charles, B. M. & Williams, R. J. (2007). *Death and Taxes: the Archaeology of a Middle Saxon Estate Centre at Higham Ferrers, Northamptonshire*. (Oxford: Oxford Archaeology).

Hardy, A., Dodd, A. & Keevill, G. D. (2003). *Ælfric's Abbey: Excavations at Eynsham Abbey*. (Oxford: Oxford Archaeology).

Haslam, J., Biek, L. & Tylecote, R.F. (1980). A middle Saxon iron smelting site at Ramsbury, Wiltshire. *Medieval Archaeology* 24, 1–68.

Havis, R. & Brooks, H. (2004). *Excavations at Stansted Airport, 1986–91*. East Anglian Archaeology 117. (Chelmsford: Essex County Council).

Heighway, C. M., Garrod, A. P. & Vince, A. G. (1979). Excavations at 1 Westgate Street, Gloucester, 1975. *Medieval Archaeology* 23, 159–213.

Henning, J. (2014). Did the 'agricultural revolution' go east with Carolingian conquest? Some reflections on early medieval rural economics of the Baiuvarii and Thuringi. In: Fries-Knoblach, J., Steuer, H., & Hines, J. eds. *Baiuvarii and Thuringi: An Ethnographic Perspective*. (Woodbridge: Boydell Press), 331–359.

Hey, G. (2004). *Yarnton: Saxon and Medieval Settlement and Landscape. Results of Excavations 1990–96*. (Oxford: Oxford Archaeology).

Higbee, L. (2009). *Hartismere High School, Eye, Suffolk: Animal Bone Assessment*. (unpublished manuscript from J. Caruth, Suffolk County Council Archaeological Service).

Higham, N. J. & Ryan, M. J. (2013). *The Anglo-Saxon World*. (New Haven & London: Yale University Press).

Hill, D. (2000). Sulh – the Anglo-Saxon plough c. 1000 AD. *Landscape History* 22, 7–19.

Hillman, G. (1981). Reconstructing crop husbandry practices from charred remains of crops. In: Mercer, R. ed. *Farming Practice in British Prehistory*. (Edinburgh: Edinburgh University Press), 123–162.

Hodges, R. (1989). *The Anglo-Saxon Achievement: Archaeology and the Beginnings of English Society*. (London: Duckworth).

Holbrook, N. & Thomas, A. (1996). The Roman and early Anglo-Saxon settlement at Wantage, Oxfordshire: excavations at Mill Street, 1993–4. *Oxoniensia* 61, 109–179.

Holmes, M. (2014). *Animals in Saxon and Scandinavian England: Backbones of Economy and Society*. (Leiden: Sidestone Press).

Hood, A. (2007). *Alpha Park, Great North Road, East Socon, Cambridgeshire: Archaeological Strip, Map and Sample: Post Excavation Assessment*. (unpublished report, Foundations Archaeology).

Hoskins, W. G. (1955). *The Making of the English Landscape*. (London: Hodder & Stoughton).

Hubbard, R. N. L. (1980). Development of agriculture in Europe and the Near East: evidence from quantitative studies. *Economic Botany* 34(1), 51–67.

Hughes, M. (1984). Rural settlement and landscape in late Saxon Hampshire. In: Faull, M. ed. *Studies in Late Anglo-Saxon Settlement*. (Oxford: Oxford University Department for External Studies), 65–80.

Hughes, M. K. & Diaz, H. F. (1994). Was there a 'Medieval Warm Period', and if so, where and when? *Climatic Change* 26, 109–142.

Hunter Blair, P. (1963). *Roman Britain and Early England 55 BC–871 AD*. (London: Thomas Nelson).

Hunter Blair, P. (1977). *An Introduction to Anglo-Saxon England*. 2nd ed. (Cambridge: Cambridge University Press).

Ivens, R. J., Busby, P. & Shepherd, N. (1995). *Tattenhoe & Westbury. Two Deserted Medieval Settlements in Milton Keynes*. (Aylesbury: Buckinghamshire Archaeological Society).

Jackson, R. P. J. & Potter, T. W. (1996). *Excavations at Stonea, Cambridgeshire 1980–85*. (London: British Museum).

Jacomet, S. (2006). *Identification of Cereal Remains from Archaeological Sites*. 2nd ed. (Basel: Basel University).

Johnstone, C. & Albarella, U. (2002). *The Late Iron Age and Romano-British Mammal and Bird Bone Assemblage from Elms Farm, Heybridge, Essex*. English Heritage Centre for Archaeology Report 45/2002. (Portsmouth: English Heritage)

Jones, A. K. G. (1975). *Wicken Bonhunt Plant Remains. A Preliminary Report*. Ancient Monuments Laboratory Report 1760. (London: English Heritage).

Jones, G. (1984). Interpretation of archaeological plant remains: ethnographic models from Greece. In: van Zeist, W. & Casparie, W. A. eds. *Plants and Ancient Man*. (Rotterdam: Balkema), 43–61.

Jones, G. (1987). A statistical approach to the identification of crop processing. *Journal of Archaeological Science* 14(3), 311–323.

Jones, G. (1991). Numerical analysis in archaeobotany. In: van Zeist, W., Wasylikowa, K., & Behre, K. E. eds. *Progress in Old World Palaeoethnobotany*. (Rotterdam: Balkema), 63–80.

Jones, M. (1981). The development of crop husbandry. In: Jones, M. & Dimbleby, G. eds. *The Environment of Man: the Iron Age to the Anglo-Saxon Period*. British Archaeological Report 87. (Oxford: British Archaeological Reports), 95–127.

Jones, M. (2009). Dormancy and the plough: Weed seed biology as an indicator of agrarian change in the first millennium AD. In: Fairbairn, A. & Weiss, E. eds. *From Foragers to Farmers. Papers in Honour of Gordon C. Hillman*. (Oxford: Oxbow Books), 58–63.

Kay, Q. O. N. (1971). *Anthemis Cotula* L. *Journal of Ecology* 59(2), 623–636.

Kenney, S. (2002). *Roman, Saxon and Medieval Occupation at the site of the former Red, White and Blue Public House, Chiefs Street, Ely*. (unpublished report, Cambridgeshire County Council).

Kenyon, D. & Watts, M. (2006). An Anglo-Saxon enclosure at Copsehill Road, Lower Slaughter: excavations in 1999. *Transactions of the Bristol and Gloucestershire Archaeological Society* 124, 73–109.

Lambrick, G. (2010). *Neolithic to Saxon Social and Environmental Change at Mount Farm, Berinsfield, Dorchester-on-Thames*. (Oxford: Oxford Archaeology).

Lavender, N. J. (1998). A Saxon building at Chadwell St. Mary: excavations at Chadwell St. Mary County Primary School 1996. *Essex Archaeology and History* 29, 48–58.

Lawson, A. J. (1983). *The Archaeology of Witton, near North Walsham, Norfolk*. East Anglian Archaeology 18. (Dereham: Norfolk Museums Service).

Lebecq, S. (2000). The role of monasteries in the systems of production and exchange of the Frankish world between the seventh and the beginning of the ninth centuries. In: Hansen, I. L. & Wickham, C. eds. *The Long Eighth Century*. (Leiden: Brill), 121–148.

Lenehan, J. J. (1986). *Grain Drying and Storage. Principles of Drying and Storing Combinable Crops*. (Dublin: An Forus Talúntais).

Lewis, C., Mitchell-Fox, P. & Dyer, C. (1997). *Village, Hamlet and Field. Changing Medieval Settlements in Central England*. (Manchester: Manchester University Press).

Losco-Bradley, S. & Kinsley, G. (2002). *Catholme. An Anglo-Saxon Settlement on the Trent Gravels in Staffordshire*. (Nottingham: Trent and Peak Archaeological Unit).

Lovell, J., Timby, J., Wakeham, G. & Allen, M. J. (2007). Iron-Age to Saxon farming settlement at Bishop's Cleeve, Gloucestershire: excavations south of Church Road, 1998 and 2004. *Transactions of the Bristol and Gloucestershire Archaeological Society* 125, 95–129.

Loveluck, C. (2007). *Rural Settlement, Lifestyles and Social Change in the Later First Millennium AD: Anglo-Saxon Flixborough in its Wider Context*. Excavations at Flixborough 4. (Oxford: Oxbow Books).

Lowe, C. (2006). *Excavations at Hoddom, Dumfriesshire: An Early Ecclesiastical Site in South-West Scotland*. (Edinburgh: society of Antiquaries of Scotland).

Lowerre, A. G., Lyons, E. R., Roberts, B. K. & Wrathmell, S. (2011). The Atlas of Rural Settlement in England GIS: Data, Metadata and Documentation [computer file]. (Swindon: English Heritage).

Lucy, S., Tipper, J. & Dickens, A. (2009). *The Anglo-Saxon Settlement and Cemetery at Bloodmoor Hill, Carlton Colville, Suffolk*. East Anglian Archaeology 131. (Cambridge: Cambridge Archaeological Unit).

Lupoi, M. (2007). *The Origins of the European Legal Order*. English tr. (Cambridge: Cambridge University Press).

Lyons, A. (2011). *Life and Afterlife at Duxford, Cambridgeshire: Archaeology and History in a Chalkland Community*. (Bar Hill: Oxford Archaeology East).

Mabey, R. (2010). *Weeds: How Vagabond Plants Gatecrashed Civilisation and Changed the Way we Think About Nature*. (London: Profile Books).

MacGowan, K. (1996). Barking Abbey. *Current Archaeology* 149, 172–178.

Mackreth, D. F. (1996). *Orton Hall Farm: A Roman and Early Anglo-Saxon Farmstead*. East Anglian Archaeology 76. (Manchester: University of Manchester & Nene Valley Archaeological Trust).

Maitland, F.W. (1897). *Domesday Book and Beyond. Three Essays in the Early History of England*. (Cambridge: Cambridge University Press).

Masefield, R. (2008). *Prehistoric and Later Settlement and Landscape from Chiltern Scarp to Aylesbury Vale: the Archaeology of the Aston Clinton Bypass, Buckinghamshire*. British Archaeological Report 473. (Oxford: Archaeopress).

Mattingly, D. (2006). *An Imperial Possession: Britain in the Roman Empire, 54 BC–AD 409*. (London: Allen Lane).

McCormick, F., Kerr, T., McClatchie, M. & O'Sullivan, A. (2014). *Early medieval agriculture, livestock and cereal production in Ireland, AD 400–1100*. British Archaeological Report S2647 (Oxford: Archaeopress).

McKerracher, M. (2014a). Agricultural development in mid Saxon England. (unpublished PhD thesis, University of Oxford).

McKerracher, M. (2014b). Landscapes of production in mid Saxon England: the monumental grain ovens. *Medieval Settlement Research* 29, 82–85.

McKerracher, M. (2016a). Saving the bacon? Reflections on the Anglo-Saxon pig. *Association for Environmental Archaeology Newsletter* (134), 4–9.

McKerracher, M. (2016b). Bread and surpluses: the Anglo-Saxon 'bread wheat thesis' reconsidered. *Environmental Archaeology* 21(1), 88–102.

Mellor, M. (1994). A synthesis of middle and late Saxon pottery, medieval and early post-medieval pottery in the Oxfordshire Region. *Oxoniensia* 59, 17–217.

Metcalf, D. (1977). Monetary affairs in Mercia in the time of Æthelbald. In: Dornier, A. ed. *Mercian Studies*. (Leicester: Leicester University Press), 87–106.

Metcalf, D. (2003). Variations in the composition of the currency at different places in England. In: Pestell, T. & Ulmschneider, K. eds. *Markets in Early Medieval Europe. Trading and 'Productive' Sites, 650–850*. (Macclesfield: Windgather), 37–47.

Miles, D. (1986). *Archaeology at Barton Court Farm, Abingdon, Oxon. An Investigation of Late Neolithic, Iron Age, Romano-British, and Saxon Settlements*. Council for British Archaeology Research Report 50. (Oxford & London: Oxford Archaeological Unit & the Council for British Archaeology).

Moffett, L. (1991). The archaeobotanical evidence for free-threshing tetraploid wheat in Britain. In: Hajnalová, E. ed. *Palaeoethnobotany and Archaeology. International Work-Group for Palaeoethnobotany 8th Symposium, Nitra-Nové Vozokany 1989*. (Nitra: Archaeological Institute of the Slovak Academy of Sciences), 233–243.

Moffett, L. (1994). Charred cereals from some ovens/kilns in late Saxon Stafford and the botanical evidence for the pre-burh economy. In: Rackham, J. ed. *Environment and Economy in Anglo-Saxon England*. Council for British Archaeology Research Report 89. (York: Council for British Archaeology), 55–64.

Moffett, L. (2006). The archaeology of medieval plant foods. In: Woolgar, C. M., Serjeantson, D., & Waldron, T. eds. *Food in Medieval England*. (Oxford: Oxford University Press), 41–55.

Moffett, L. (2011). Food plants on archaeological sites: the nature of the archaeobotanical record. In: Hamerow, H., Hinton, D. A., & Crawford, S. eds. *The Oxford Handbook of Anglo-Saxon Archaeology*. (Oxford: Oxford University Press), 346–360.

Moore, J. (2001). Excavations at Oxford Science Park, Littlemore, Oxford. *Oxoniensia* 66, 163–219.

Mortimer, R. (1996). *Excavations of a Group of Anglo-Saxon Features at Denny End, Waterbeach, Cambridgeshire*. (unpublished report, Cambridge Archaeological Unit).

Mortimer, R. (1998). *Excavation of the Middle Saxon to Medieval Village at Lordship Lane, Cottenham, Cambridgeshire*. (unpublished report, Cambridge Archaeological Unit).

Mortimer, R. (2000). Village development and ceramic sequence: the middle to late Saxon village at Lordship Lane, Cottenham, Cambridgeshire. *Proceedings of the Cambridge Antiquarian Society* 89, 5–33.

Mortimer, R. (2003). *Rosemary Lane, Church End, Cherry Hinton*. (unpublished report, Cambridge Archaeological Unit).

Mortimer, R. & Evans, C. (1996). *An Archaeological Excavation at Hinxton Quarry, Cambridgeshire, 1995*. (unpublished report, Cambridge Archaeological Unit).

Mortimer, R., Regan, R. & Lucy, S. (2005). *The Saxon and Medieval Settlement at West Fen Road, Ely: The Ashwell Site*. East anglian Archaeology 110. (Cambridge: Cambridge Archaeological Unit).

Mudd, A. (2002). *Excavations at Melford Meadows, Brettenham, 1994: Romano-British and Early Saxon Occupations*. (Oxford: Oxford Archaeological Unit).

Mudd, A. (2007). *Bronze Age, Roman and Later Occupation at Chieveley, West Berkshire. The archaeology of the A34/M4 Road Junction Improvement*. (Oxford: British Archaeological Reports).

Mudd, A. & Webster, M. (2011). *Iron Age and Middle Saxon Settlements at West Fen Road, Ely, Cambridgeshire: The Consortium Site*. British Archaeological Report 538. (Oxford: Archaeopress).

Murphy, P. (1994). The Anglo-Saxon landscape and rural economy: some results from sites in East Anglia and Essex. In: Rackham, J. ed. *Environment and Economy in Anglo-Saxon England*. Council for British Archaeology Research Report 89. (York: Council for British Archaeology), 23–39.

Murphy, P. (2010). The landscape and economy of the Anglo-Saxon coast: new archaeological evidence. In: Higham, N. J. & Ryan, M. J. eds. *Landscape Archaeology of Anglo-Saxon England*. (Woodbridge: Boydell Press), 211–221.

Murphy, P. & Fryer, V. (2009). Staunch Meadow, Brandon. Valley sediments and plant macrofossils. (unpublished manuscript from author).

Murray, J. (2005). Excavations at Station Road, Gamlingay, Cambridgeshire. *Anglo-Saxon Studies in Archaeology and History* 13, 173–330.

Mynard, D. C. & Zeepvat, R. J. (1992). *Excavations at Great Linford, 1974–80*. (Aylesbury: Buckinghamshire Archaeological Society).

Naylor, J. (2012). Coinage, trade and the origins of the English Emporia, ca. AD 650–750. In: Gelichi, S. & Hodges, R. eds. *From One Sea to Another. Trading Places in the European and Mediterranean Early Middle Ages*. (Turnhout: Brepols), 237–266.

Nesbitt, M. & Samuel, D. (1996). From staple crop to extinction? The archaeology and history of the hulled wheats. In: Padulosi, S., Hammer, K., & Heller, J. eds. *Hulled Wheats. Promoting the Conservation and Use of Underutilized and Neglected Crops*. (Rome: International Plant Genetic Resources Institute), 40–99.

Newman, J. (1992). The late Roman and Anglo-Saxon settlement pattern in the Sandlings of Suffolk. In: Carver, M. ed. *The Age of Sutton Hoo: The Seventh Century in North-Western Europe*. (Woodbridge: Boydell Press), 25–38.

Newton, A. A. S. (2010). *Saxon and Medieval Settlement at The Old Bell, Marham, Norfolk. Research Archive Report*. (unpublished report, Archaeological Solutions).

Noddle, B. (1975). A comparison of the animal bones from 8 medieval sites in Southern Britain. In: Clason, A. T. ed. *Archaeozoological Studies*. (Amsterdam: American Elsevier), 248–260.

O'Brien, L. (2016). *Bronze Age Barrow, Early to Middle Iron Age Settlement and Burials, and Early Anglo-Saxon Settlement at Harston Mill, Cambridgeshire*. (Bury St Edmunds: Archaeological Solutions).

O'Connor, T. (2000). *The Archaeology of Animal Bones*. (Stroud: Sutton).

O'Connor, T. (2011). Animal husbandry. In: Hamerow, H., Hinton, D. A., & Crawford, S. eds. *The Oxford Handbook of*

Anglo-Saxon Archaeology. (Oxford: Oxford University Press), 361–376.

O'Sullivan, D. (2001). Space, silence and shortage on Lindisfarne. The archaeology of asceticism. In: Hamerow, H. & Macgregor, A. eds. *Image and Power in the Archaeology of Early Medieval Britain*. (Oxford), 33–52.

Oosthuizen, S. (2005). New light on the origins of open-field farming? *Medieval Archaeology* 49, 165–193.

Oosthuizen, S. (2013a). *Tradition and Transformation in Anglo-Saxon England: Archaeology, Common Rights and Landscape*. (London: Bloomsbury Academic).

Oosthuizen, S. (2013b). Debate: the Emperor's old clothes and the origins of medieval nucleated settlements and their open fields. *Medieval Settlement Research* 28, 96–98.

Pals, J. P. & van Dierendonck, M. C. (1988). Between flax and fabric: cultivation and processing of flax in a medieval peat reclamation settlement near Midwoud (Prov. Noord Holland). *Journal of Archaeological Science* 15, 237–251.

Patrick, C. & Rátkai, S. (2011). Hillside Meadow, Fordham. In: Cuttler, R., Martin-Bacon, H., Nichol, K., Patrick, C., Perrin, R., Rátkai, S., Smith, M., & Williams, J. eds. *Five Sites in Cambridgeshire. Excavations at Woodhurst, Fordham, Soham, Buckden and St. Neots, 1998–2002*. British Archaeological Report 528. (Oxford: Archaeopress), 41–123.

Payne, S. (1973). Kill-off patterns in sheep and goats: the mandibles from Aşvan Kale. *Anatolian Studies* 23, 281–303.

Payne, S. (1999). Animal husbandry. In: Lapidge, M., Blair, J., Keynes, S., & Scragg, D. eds. *The Blackwell Encyclopaedia of Anglo-Saxon England*. (Oxford: Blackwell), 38–39.

Pelling, R. (2003). Early Saxon cultivation of emmer wheat in the Thames Valley and its cultural implications. In: Robson Brown, K. A. ed. *Archaeological Sciences 1999. Proceedings of the Archaeological Sciences Conference, University of Bristol, 1999*. British Archaeological Report S1111. (Oxford: Archaeopress), 103–110.

Pelling, R. & Robinson, M. (2000). Saxon emmer wheat from the Upper and Middle Thames Valley, England. *Environmental Archaeology* 5, 117–119.

Percival, J. (2001). *Downham Market Excavation Areas 4 & 5 (Land North of A1122 Downham Market Bypass)*. (unpublished report, Norfolk Archaeological Unit).

Pestell, T. (2011). Markets, emporia, wics, and 'productive' sites: pre-Viking trade centres in Anglo-Saxon England. In: Hamerow, H., Hinton, D. A., & Crawford, S. eds. *The Oxford Handbook of Anglo-Saxon Archaeology*. (Oxford: Oxford University Press), 556–579.

Phillips, M. (2005). Excavations of an Early Saxon settlement at Pitstone. *Records of Buckinghamshire* 45, 1–32.

Pine, J. (2001). The excavation of a Saxon settlement at Cadley Road, Collingbourne Ducis, Wiltshire. *Wiltshire Archaeological and Natural History Magazine* 94, 88–117.

Pine, J. (2009). *Latton Quarry, Latton, Wiltshire. A Post-Excavation Assessment for Hills Quarry Products*. (unpublished report, Thames Valley Archaeological Services).

Pine, J. & Ford, S. (2003). Excavation of Neolithic, Late Bronze Age, Early Iron Age and Early Saxon Features at St. Helen's Avenue, Benson, Oxfordshire. *Oxoniensia* 68, 131–178.

Pitts, M. (2011). Kent plough find challenges farming history. *British Archaeology* 118, 7.

Pollard, J. (1996). *Excavations at Bourn Bridge, Pampisford, Cambridgeshire: Part 2, Roman and Saxon*. (unpublished report, Cambridge Archaeological Unit).

Powell, A. (2011). *An Iron Age Enclosure and Romano-British Features at High Post, near Salisbury*. (Salisbury: Wessex Archaeology).

Powlesland, D. (2011). Archaeological Excavations in Sherburn, Vale of Pickering, North Yorkshire. (The Landscape Research Centre: <http://www.landscaperesearchcentre.org/ArchaeologicalExcavationsinSherburn2011.pdf>, accessed October 2013).

Preston, S. (2007). Bronze Age occupation and Saxon features at the Wolverton Turn Enclosure, near Stony Stratford, Milton Keynes: investigations by Tim Schadla-Hall, Philip Carstairs, Jo Lawson, Hugh Beamish, Andrew Hunn, Ben Ford and Tess Durden, 1972 to 1994. *Records of Buckinghamshire* 47(1), 81–117.

Price, E. (2000). *Frocester. A Romano-British Settlement, its Antecedents and Successors*. (Stonehouse: Gloucester and District Archaeological Research Group).

Pryor, F. (1996). Sheep, stockyards and field systems: Bronze Age livestock populations in the Fenlands of eastern England. *Antiquity* 70(268), 313–324. Available at: http://www.scopus.com/inward/record.url?eid=2-s2.0-0030478278&partnerID=40&md5=c17743d983ecd3f2941e5a3e9701c87a.

Pryor, F. (2010). *The Making of the British Landscape*. (London: Penguin Books).

Purcell, N. (1995). The Roman villa and the landscape of production. In: Cornell, T. J. & Lomas, K. eds. *Urban Society in Roman Italy*. (London: University College London), 151–179.

Puy, A. & Balbo, A. L. (2013). The genesis of irrigated terraces in al-Andalus. A geoarchaeological perspective on intensive agriculture in semi-arid environments (Ricote, Murcia, Spain). *Journal of Arid Environments* 89, 45–56.

Rackham, O. (1986). *The History of the Countryside*. (London: Dent).

Rahtz, P. & Meeson, R. (1992). *An Anglo-Saxon Watermill at Tamworth*. Council for British Archaeology Research Report 83. (York: Council for British Archaeology).

Rees, S.E. (1979). *Agricultural Implements in Prehistoric and Roman Britain* British Archaeological Report 69. (Oxford: British Archaeological Reports).

Reynolds, A. (2003). Boundaries and settlements in later sixth to eleventh century England. In: Griffiths, D., Reynolds, A., & Semple, S. eds. *Boundaries in Early Medieval Britain*. Anglo-Saxon Studies in Archaeology and History 12. (Oxford: Oxford University School of Archaeology), 98–136.

Rickett, R. (1995). *The Anglo-Saxon Cemetery at Spong Hill, North Elmham, Part VII: The Iron Age, Roman and Early Saxon Settlement*. East Anglian Archaeology 73. (Gressenhall: Norfolk Museums Service).

Rickett, R. J. (1975). *Post-Roman and Medieval Drying Kilns*. (unpublished BA thesis, University College, Cardiff).

Rippon, S. (2008). *Beyond the Medieval Village*. (Oxford: Oxford University Press).

Rippon, S. (2010). Landscape change during the 'long eighth century' in southern England. In: Higham, N. J. & Ryan, M. J. eds. *The Landscape Archaeology of Anglo-Saxon England*. (Woodbridge: Boydell Press), 39–64.

Rippon, S., Fyfe, R. M. & Brown, A. G. (2006). Beyond villages and open fields: the origins and development of a historic landscape characterised by dispersed settlement in south-west England. *Medieval Archaeology* 50, 31–70.

Rippon, S., Smart, C. & Pears, B. (2015). *The Fields of Britannia: Continuity and Change in the Late Roman and Early Medieval Landscape.* (Oxford: Oxford University Press).

Roberts, B. K. & Wrathmell, S. (2000). *An Atlas of Rural Settlement in England.* (London: English Heritage).

Robinson, M. *Assessment of Carbonised Plant Remains from Rivers Nightclub, n.d. Benson, Oxon (SAB99114).* (unpublished report on file at Oxfordshire Historic Environment Records).

Robinson, M. (1992). Environmental archaeology of the river gravels: past achievements and future directions. In: Fulford, M. & Nichols, E. eds. *Developing Landscapes of Lowland Britain. The Archaeology of the British Gravels: A Review.* (London: Society of Antiquaries of London), 47–62.

Robinson, M. (1997). *Charred Plant Remains from Walton, Aylesbury, Bucks.* (unpublished report on file at Buckinghamshire Historic Environment Records).

Robinson, M. (2011). The paleoecology of alluvial hay meadows in the Upper Thames valley. *Fritillary* 5, 47–57.

Rowley, T. & Brown, L. (1981). Excavations at Beech House Hotel, Dorchester-on-Thames 1972. *Oxoniensia* 46, 1–55.

Rowley, T. & Steiner, M. (1996). *Cogges Manor Farm Witney Oxfordshire. The Excavations from 1986–1994 and the Historic Building Analysis.* (Oxford: University of Oxford & Oxfordshire County Council).

Ruas, M.-P. (2005). Aspects of early medieval farming from sites in Mediterranean France. *Vegetation History and Archaeobotany* 14, 400–415.

Sadler, P. (1998). *Animal Remains from Croft Road, Walton Road Stores and Walton Lodge Lane.* (unpublished report on file at Buckinghamshire Historic Environment Records).

Scull, C. (2009). *Early Medieval (Late 5th–Early 8th Centuries AD) Cemeteries at Boss Hall and Buttermarket, Ipswich, Suffolk.* Society for Medieval Archaeology Monograph 27. (Leeds: Society for Medieval Archaeology).

Shoesmith, R. (1982). *Hereford City Excavations Volume 2: Excavations on and Close to the Defences.* Council for British Archaeology Research Report 46. (London: Council for British Archaeology).

Sigaut, F. (1988). A method for identifying grain storage techniques and its application for European agricultural history. *Tools & Tillage* 6(1), 3–32.

Sims, R.E. (1978). Man and vegetation in Norfolk. In: Limbrey, S. & Evans, J.G. eds. Science in Archaeology. (London: Thames Hudson), 283–302.

Snape, M. (2003). A horizontal-wheeled watermill of the Anglo-Saxon period at Corbridge, Northumberland, and its river environment. *Archaeologia Aeliana* 32, 37–72.

Stace, C. A. (2010). *New Flora of the British Isles.* 3rd ed. (Cambridge: Cambridge University Press).

Stone, P. (2009). *82–84 Walton Street, Aylesbury, Buckinghamshire. Research Archive Report.* (unpublished report by Archaeological Solutions Ltd).

Swanton, M. (1975). *Anglo-Saxon Prose.* (London: Dent).

Sykes, N. J. (2007). *The Norman Conquest: A Zooarchaeological Perspective.* British Archaeological Report S1656 (Oxford: Archaeopresss).

Taylor, G. (2003). An early to middle Saxon settlement at Quarrington, Lincolnshire. *Antiquaries Journal* 83, 231–280.

Taylor, K. & Ford, S. (2004). Late Bronze Age, Iron Age, Roman and Saxon sites along the Oxfordshire section. In: Ford, S., Howell, I. J., & Taylor, K. eds. *The Archaeology of the Aylesbury-Chalgrove Gas Pipeline and The Orchard, Walton Road, Aylesbury.* (Reading: Thames Valley Archaeological Services), 25–58.

Tester, A. (2006). *RAF Lakenheath, New Dental Clinic ERL 101. A report on the archaeological excavations, 2001/2.* (unpublished report, Suffolk County Council Archaeological Service).

Tester, A., Anderson, S., Riddler, I. & Carr, R. (2014). *Staunch Meadow, Brandon, Suffolk: a High Status Middle Saxon Settlement on the Fen Edge.* (Bury St Edmunds: Suffolk County Council Archaeological Service).

Thirsk, J. (1964). The common fields. *Past and Present* 29, 3–25.

Thirsk, J. (1966). The origins of the common fields. *Past and Present* 33, 142–147.

Thomas, G. (2009). Uncovering an Anglo-Saxon monastery in Kent. Interim report on University of Reading excavations at Lyminge, 2008 [Online]. Available at: http://www.reading.ac.uk/archaeology/research/Projects/arch_Lyminge.aspx.

Thomas, G. (2010). *The Later Anglo-Saxon Settlement at Bishopstone: A Downland Manor in the Making.* Council for British Archaeology Research Report 163 (York: Council for British Archaeology).

Thomas, G. (2011). Overview: craft production and technology. In: Hamerow, H., Hinton, D. A., & Crawford, S. eds. *The Oxford Handbook of Anglo-Saxon Archaeology.* (Oxford: Oxford University Press), 405–422.

Thomas, G. (2013). Life before the minster: the social dynamics of monastic foundation at Anglo-Saxon Lyminge, Kent. *Antiquaries Journal* 93, 109–145.

Thomas, G., McDonnell, G., Merkel, J. & Marshall, P. (2016). Technology, ritual and Anglo-Saxon agrarian production: the biography of a seventh-century plough coulter from Lyminge, Kent. *Antiquity* 90(351), 742–758.

Tipper, J. (2004). *The Grubenhaus in Anglo-Saxon England: an Analysis and Interpretation of the Evidence from a Most Distinctive Building Type.* (Yedingham: Landscape Research Centre).

Tipper, J. (2007). *West Stow, Lackford Bridge Quarry (WSW 030). A report on a rescue excavation undertaken in 1978–9.* (unpublished report, Suffolk County Council Archaeological Service).

Trimble, G. L. (2001). *Report on an Archaeological Excavation at Land off London Road, Downham Market, Norfolk, Areas 1–3.* (unpublished report, Norfolk Archaeological Unit).

Troldtoft Andresen, S. & Karg, S. (2011). Retting pits for textile fibre plants at Danish prehistoric sites dated between 800 BC and AD 1050. *Vegetation History and Archaeobotany* 20, 517–526.

Ulmschneider, K. (2000). *Markets, Minsters, and Metal-Detectors. The Archaeology of Middle Saxon Lincolnshire and Hampshire Compared.* British Archaeological Report 307. (Oxford: Archaeopress).

Ulmschneider, K. (2011). Settlement hierarchy. In: Hamerow, H., Hinton, D. A., & Crawford, S. eds. *The Oxford Handbook of Anglo-Saxon Archaeology.* (Oxford: Oxford University Press), 156–171.

Upex, S. G. (2002). Landscape continuity and the fossilization of Roman fields. *Archaeological Journal* 159, 77–108.

Vaughan-Williams, A. (2005). Report on the plant remains from Forbury House Reading. In: Edwards, C. & Adams, S. eds. *Forbury House, Reading, Berkshire. Archive Report.* (unpublished report, Berkshire Historic Environment Record: AOC Archaeology Group), 44–48.

van der Veen, M. (1989). Charred grain assemblages from Roman-period corn driers in Britain. *Archaeological Journal* 146, 302–319.

van der Veen, M. (2007). Formation processes of desiccated and carbonized plant remains - the identification of routine practice. *Journal of Archaeological Science* 34, 968–990.

van der Veen, M. & Jones, G. (2006). A re-analysis of agricultural production and consumption: implications for understanding the British Iron Age. *Vegetation History and Archaeobotany* 15, 217–228.

van der Veen, M. & O'Connor, T. (1998). The expansion of agricultural production in late Iron Age and Roman Britain. In: Bayley, J. ed. *Science in Archaeology: An Agenda for the Future.* (London: English Heritage), 127–143.

Wade-Martins, P. (1980). *Excavations in North Elmham Park 1967–1972.* East Anglian Archaeology 9. (Gressenhall: Norfolk Museums Service).

Wade, K. (1980). A settlement site at Bonhunt Farm, Wicken Bonhunt, Essex. In: Buckley, D. G. ed. *Archaeology in Essex to AD 1500.* Council for British Archaeology Research Report 34. (London: Council for British Archaeology), 96–102.

Wall, W. (2011). Middle Saxon Iron Smelting near Bonemills Farm, Wittering, Cambridgeshire. *Anglo-Saxon Studies in Archaeology and History* 17, 87–100.

Wallis, S. & Waughman, M. (1998). *Archaeology and the Landscape in the Lower Blackwater Valley.* East Anglian Archaeology 82. (Chelmsford: Essex County Council).

Waton, P. V. (1982). Man's impact on the chalklands: some new pollen evidence. In: Bell, M. & Limbrey, S. eds. *Archaeological Aspects of Woodland Ecology.* British Archaeological Report S146. (Oxford: British Archaeological Reports), 75–91.

Watts, M. (2002). *The Archaeology of Mills and Milling.* (Stroud: Tempus).

Webster, G., Fowler, P., Noddle, B. & Smith, L. (1985). The excavation of a Romano-British rural establishment at Barnsley Park, Gloucestershire, 1961–1979. Part III. *Transactions of the Bristol and Gloucestershire Archaeological Society* 103, 73–100.

Webster, L. (2012). *Anglo-Saxon Art: A New History.* (London: British Museum Press).

Welch, M. (2011). The Mid Saxon 'Final Phase'. In: Hamerow, H., Hinton, D. A., & Crawford, S. eds. *The Oxford Handbook of Anglo-Saxon Archaeology.* (Oxford: Oxford University Press), 266–287.

West, S. E. (1985). *West Stow. The Anglo-Saxon Village.* East Anglian Archaeology 24. (Ipswich: Suffolk County Council).

Whitelock, D. (1952). *The Beginnings of English Society.* (Harmondsworth: Penguin Books).

Whitelock, D. (1979). *English Historical Documents, Vol. I: 500–1042.* 2nd ed. (London: Methuen).

Wickham, C. (2000). Introduction. In: Hansen, I. & Wickham, C. eds. *The Long Eighth Century.* (Leiden: Brill), ix–x.

Wickham, C. (2009). *The Inheritance of Rome. A History of Europe from 400 to 1000.* (London: Allen Lane).

Williams, G. (2008). *An Archaeological Evaluation at Milton Park, Didcot, Oxfordshire.* (unpublished report, John Moore Heritage Services).

Williams, P. & Newman, R. (2006). *Market Lavington, Wiltshire, An Anglo-Saxon Cemetery and Settlement. Excavations at Grove Farm, 1986–90.* Wessex Archaeology Report 19. (Salisbury: Wessex Archaeology).

Williams, R. J. (1993). *Pennyland and Hartigans. Two Iron Age and Saxon sites in Milton Keynes.* (Aylesbury: Buckinghamshire Archaeological Society).

Williams, R. J., Hart, P. J. & Williams, A. T. L. (1996). *Wavendon Gate. A Late Iron Age and Roman settlement in Milton Keynes.* (Aylesbury: Buckinghamshire Archaeological Society).

Williams, R. J. & Zeepvat, R. J. (1994). *Bancroft. A Late Bronze Age/Iron Age Settlement, Roman Villa and Temple-Mausoleum.* (Aylesbury: Buckinghamshire Archaeological Society).

Williamson, T. (2003). *Shaping Medieval Landscapes.* (Macclesfield: Windgather).

Williamson, T. (2013). *Environment, Society and Landscape in Early Medieval England: Time and Topography.* (Woodbridge: Boydell Press).

Wilson, A. (2007). The uptake of mechanical technology in the ancient world: the water-mill [Online]. Available at: http://oxrep.classics.ox.ac.uk/ [Accessed: 8 August 2012].

Wilson, D. M. & Hurst, J. G. (1958). Medieval Britain in 1957. *Medieval Archaeology* 2, 183–213.

Wiltshire, P. E. J. (1988). *Microscopic Analysis of Sediments Taken from the Edge of Micklemere, Pakenham, Suffolk.* (Ancient Monuments Laboratory Report 209/88).

Wiltshire, P.E.J. (1990). *A palynological analysis of sediments from Staunch Meadow, Brandon, Suffolk.* Ancient Monuments Laboratory Report 73/90. (London: English Heritage).

Wood, M. (2010). *The Story of England.* (London: Penguin Books).

Woodward, A. & Leach, P. (1993). *The Uley Shrines. Excavation of a Ritual Complex on West Hill, Uley, Gloucestershire: 1977–9.* (London: English Heritage).

Wormald, P. (1984). *Bede and the Conversion of England: the Charter Evidence.* (Jarrow: Jarrow Trust).

Wright, D. W. (2015). Early medieval settlement and social power: the middle Anglo-Saxon 'home farm'. *Medieval Archaeology* 59, 24–46.

Wright, J. (2004). *Anglo-Saxon Settlement at Cherry Orton Road, Orton Waterville, Peterborough. Report on the 2003 Archaeological Excavations.* (unpublished report, Wessex Archaeology).

Wright, J., Leivers, M., Seager Smith, R. & Stevens, C. J. (2009). *Cambourne New Settlement. Iron Age and Romano-British settlement on the clay uplands of west Cambridgeshire.* (Salisbury: Wessex Archaeology).

Zeepvat, R. J. (1993). The Milton Keynes Project. *Records of Buckinghamshire* 33, 49–63.

Zimmermann, W. H. (1992). The 'Helm' in England, Wales, Scandinavia and North America. *Vernacular Architecture* 23, 34–43.